INDUSTRIALIZATION OF DRUG DISCOVERY

Drug Discovery Series

Series Editor

Andrew Carmen

Johnson & Johnson PRD, LLC
San Diego, California, U.S.A.

1. Virtual Screening in Drug Discovery, *edited by Juan Alvarez and Brian Shoichet*
2. Industrialization of Drug Discovery: From Target Selection Through Lead Optimization, *edited by Jeffrey S. Handen, Ph.D.*
3. Phage Display in Biotechnology and Drug Discovery, *edited by Sachdev S. Sidhu*
4. G Protein-Coupled Receptors in Drug Discovery, *edited by Kenneth H. Lundstrom and Mark L. Chiu*

INDUSTRIALIZATION OF DRUG DISCOVERY

From Target Selection Through Lead Optimization

EDITED BY

JEFFREY S. HANDEN

Taylor & Francis
Taylor & Francis Group

Boca Raton London New York Singapore

A CRC title, part of the Taylor & Francis imprint, a member of the Taylor & Francis Group, the academic division of T&F Informa plc.

Published in 2005 by
CRC Press
Taylor & Francis Group
6000 Broken Sound Parkway NW, Suite 300
Boca Raton, FL 33487-2742

Printed in the United States of America on acid-free paper
10 9 8 7 6 5 4 3 2 1

International Standard Book Number-10: 0-8247-2391-0 (Hardcover)
International Standard Book Number-13: 978-0-8247-2391-0 (Hardcover)
Library of Congress Card Number 2004063486

Library of Congress Cataloging-in-Publication Data

Industrialization of drug discovery : from target selection through lead optimization / edited by Jeffrey S. Handen.
p. cm. (Drug Discovery series)
Includes bibliographical references and index.
ISBN 0-8247-2391-0 (alk. paper)
1. Drug development. 2. Pharmaceutical industry. I. Handen, Jeffrey S. II. Title.

RM301.25.I53 2005
615'.19--dc22 2004063486

T&F informa
Taylor & Francis Group
is the Academic Division of T&F Informa plc.

Visit the Taylor & Francis Web site at
http://www.taylorandfrancis.com

and the CRC Press Web site at
http://www.crcpress.com

Dedication

Dedicated to my family, Connie, Alex, and Max Handen

Preface

From Alexander Fleming's pioneering work with the discovery of penicillin to the no less monumental technical feat of mapping the human genome, the latter half of the 20th century held the promise of the development of drugs and other therapeutics for the final eradication of bacterial infectious diseases and the opening of the door to biotechnology-based medicines for the treatment and cure of disease at the very molecular level. Though the past 50 years have been marked by unheralded advances in the research and development of therapeutics, the expectations of a radical revolution in the treatment of disease have yet to come to fruition. The pharmaceutical industry, as a whole, is arguably suffering from an innovation deficit as it struggles to cope with new and ever-increasing economic pressures. For the period 1989 to 2000, only 15% of the 1035 new drug applications (NDAs) submitted to the U.S. Food and Drug Administration (FDA) were judged by the agency to be new molecular entities (NMEs) worthy of clinical priority review. The biotechnology industry has fared somewhat better, positioning itself as an engine of growth for the future. Currently, there are approximately 350 biotechnology-derived drugs on the market. That number is expected to increase by 50% within the next 5 years. However, it still remains to be seen if these drugs are better, faster, safer, and truly represent a new paradigm. What is clear, from numerous oft-cited statistics, is that drug discovery and development is getting longer, more expensive, and no better, suffering from the same clinical attrition and safety-related market-withdrawal rates today as it did 20 years ago. Moreover, the explosion of potential new targets from all the various *-omics* initiatives and endeavors, and the hunt for that elusive needle in a haystack, which now characterizes drug discovery and development, will only get more costly and more lengthy. For the full promises of the last 50 years to be realized, a new approach to drug discovery and development must be initiated. That approach can perhaps be best characterized as a modern-day back-to-the-basics approach—the intent being to understand the science, unravel the story, and then intelligently apply technology to bring to bear the weapons of industrialization, not just automation, in order to start with a bigger needle and a smaller haystack.

The aim of this book is to jointly explore such concepts, examine the current state of the art, look at some early success stories, and initiate a dialogue and challenge to the status quo. The goal is more specific targets, more selective compounds, and less time.

This book consists of 10 chapters, briefly summarized below. Chapter 1, "Drug Discovery in the Modern Age: How We Got Here and What Does It Mean?" introduces the historical background and current status of drug discovery. Readers will be able to understand the basic concepts and tenets underlying modern drug discovery, how they have evolved, and various approaches and strategies to modern drug discovery.

Chapter 2, "The Regulatory Age," examines the number, import, and complexity of FDA regulations, requirements, guidelines, and draft documents so one can see the underlying principles that guide so many FDA actions. There are three underlying precepts that together provide a foundation for and an understanding of all the myriad FDA actions: (1) proximal causality, (2) risk assessment, and (3) self-regulation. These three principles can be used heuristically as well, providing a basis for the prediction of future FDA actions, and a projection of interpretations and emphases. Since the FDA's inception, these three principles have governed the ebb and flow of debate between proponents of the industry promotion and protection (providing more medical options) and the proponents of public health and safety (providing fewer but safer options).

Chapter 3, "Industrialization, Not Automation," discusses the productivity gap in the pipelines of the pharmaceutical industry—this despite the promised returns from large investments in the mid-1990s in high-throughput technologies for screening millions of compounds against the large numbers of new targets from genomics and later the human genome project. As the health care environment becomes increasingly difficult, and organizations become increasingly large and complex, most are struggling to maintain their current level of productivity. However, recognizing the need for greater productivity, pharmaceutical executives are demanding significantly greater throughput from existing resources. Simply increasing the level of investment in technology, particularly process automation, will not deliver this. Success will be achieved through properly planned and implemented process industrialization in an innovative R&D-based environment. This chapter highlights the key elements of industrialization that have not changed since the industrial revolution in England in the 18th century. It discusses how industrialization has evolved and developed from the second industrial revolution, through Ford's production approaches to car manufacture, to modern manufacturing engineering disciplines now standard in many high-technology industries. Finally, it will address how modern industrial approaches can be applied to pharmaceutical discovery organizations to significantly impact the productivity gap.

In Chapter 4, "Compound Management," corporate compound collections are defined as the "crown jewels" of the organization, and that has led to the need to invest considerable talent and resources in managing and utilizing this asset. Indeed, with all groups within the companies dependent on the compound management activities, and the increasing recognition that the makeup of the compound collection is of strategic importance, investment in compound management activities is increasing dramatically.

Chapter 5, "High-Throughput Screening," deals with the classical division of pharmaceutical research into four phases, commencing with target identification and ending with preclinical. Target identification (TI) is the process by which the hypothesis of the involvement of a particular molecular target is postulated. The next stage, lead identification (LI), hopefully delivers several chemical lead series that show a demonstrable effect on the disease target of interest. Today, high-throughput screening (HTS) is an integral component of the LI process. Successful LI programs result in a project's transition to lead optimization (LO), where medicinal chemistry is brought to bear on the optimization of the structure-activity relationships (SARs) around specific pharmacophore classes that were identified in LI. Finally, optimized lead compounds from LO programs enter preclinical, where their overall selectivity and toxicity profiles are assessed as a precursor for their entry into clinical development. Each of these steps is associated with a significant attrition rate for reasons of lack of efficacy, specificity, or toxicity.

Chapter 6, "Parallel Lead Optimization," introduces the selection of the preclinical candidate in exploring each promising chemical series through a hierarchical cascade of tests, defined by the required therapeutic profile consisting of multiple, sometimes interdependent, parameters. The current paradigm in the industry is that a significant improvement in both success rate and efficiency will result when multiple lead series are investigated in parallel with simultaneous optimization of all relevant parameters.

Chapter 7, "Knowledge Management," explores the large demand in organizations for people who are interested in understanding information and knowledge management within a corporate environment. Information and knowledge management are burgeoning fields in the business world. Companies realize the need to access the knowledge base located in the minds of their employees, as well as the innumerable documents and files they generate each day. A prerequisite of knowledge management is the foundational capability of accessing and exploiting information on an enterprise-wide scale. People with the ability to diagnose informational needs, retrieve relevant information, and repackage that information in the most appropriate manner are in high demand. The desired end result is to better organize internal knowledge and accessibility, leading ultimately to productivity and success.

Chapter 8, "Understanding the Value of Research," discusses the continuous investments life science companies must make in research, with the goal that these investments will help to discover new therapies for unmet medical needs. In recent years, significant advances in medical science have been achieved, such that there is an increasing wealth of investment choices available. The challenge is to select those investments that will produce the best return. Investment opportunities include drug discovery programs, discovery technologies, drug platforms, informatics systems, and research infrastructure.

Of course, more often than not there are far more investment opportunities than money available. Faced with limited resources, management must rationalize and make choices between alternative investments. To help make these choices, most investors across industries tend to measure the value, expected return, and risk of each investment opportunity. Based on these measures, the investors can make an informed decision as to which investment they prefer given their different objectives.

Chapter 9, "Collaboration in a Virtual and Global Environment," begins by defining collaboration and suggests why it is so much more important today than ever before, especially in light of the fact that almost all teams now encounter some element of being virtual, if not global as well. After describing a number of obstacles that can and do get in the way of effective collaboration, while alluding to why many of these exist, the chapter then explains how an organization should proceed, which includes an implementation approach and the components that must necessarily be addressed. It lists some of the principles, assumptions, and elements necessary for developing a collaboration strategy so the reader can understand the groundwork that must take place well before proceeding. The chapter concludes with a case study of a successful implementation of a collaborative environment in an R&D organization.

Chapter 10, "From Genome to Drug: Ethical Issues," concerns itself with the question, Why should the biopharmaceutical industry concern itself with bioethics? and explores compelling answers. These companies are investing heavily in genomics and proteomics for their potential to positively transform both drug discovery and drug development. Rightly or wrongly, DNA-based research is evaluated very differently from other types of traditional drug research. Advances in biotechnology are viewed as having the potential to shape the very nature of human life because they manipulate DNA, the building blocks of life. Descriptions by scientists of the importance of mapping the human genome have led the public to expect a transformation in biological research and medical practice. Characterization of the human genome sequence as the instruction book for human biology only reinforces the notion that biotechnology overlaps with fundamental questions about what it means to be human.

Innovations in biotechnology have the potential to not only improve our health status, but also change how we live, what we value, and who we are—leading to debates about the moral dimensions of personhood. Therefore, it is not surprising that a wide range of stakeholder groups want a role in determining how biotechnology should be used, regulated, and financed, particularly as they have argued that the current oversight mechanisms are inadequate to meet the new challenges.

Jeffrey S. Handen, Ph.D.

The Editor

Dr. Jeffrey S. Handen is employed in the area of business process improvement. He has partnered with pharmaceutical, biotech, and clinical research organization (CRO) clients to define, develop, and optimize their R&D capabilities in discovery, preclinical, and clinical development. His work has incorporated strategic change and organizational redesign as well as process improvement identification, solution development, and implementation. He has more than 15 years of experience in life sciences R&D and R&D process management, in addition to expertise in program and project management, product and technology assessment, genomics, and other advanced technologies. Dr. Handen has held positions in research and development and project management in industry, consulting, and at the University of Pennsylvania and the National Institutes of Health.

Dr. Handen received his Ph.D. in neurosciences at the George Washington University and resides in Wallingford, Pennsylvania, U.S.A. He has published numerous articles in both business and peer-reviewed scientific journals and is an oft-invited speaker at national drug discovery and development conferences.

Contributors

Richard Archer
The Automation Partnership
Cambridge, United Kingdom

John Barrett
ITI Associates
Ambler, Pennsylvania

Mark Beggs, Ph.D.
The Automation Partnership
Cambridge, United Kingdom

Alan Beresford
ADME Lead Optimisation
Inpharmatica Ltd.
Cambridge, United Kingdom

Philippe Bey
Thrasos, Inc.
Waltham, Massachusetts

Yann Bonduelle, Ph.D.
PricewaterhouseCoopers
London, United Kingdom

Beverly Buckta
AstraZeneca
Wilmington, Delaware

Paul Bussey
PricewaterhouseCoopers
London, United Kingdom

Patricia Deverka, M.D., M.S.
Medco Health Solutions
Franklin Lakes, New Jersey

Victoria Emerick
The Automation Partnership
Cambridge, United Kingdom

Steven L. Gallion, Ph.D.
Cellular Genomics, Inc.
Branford, Connecticut

Jeffrey S. Handen, Ph.D.
Merck Research Laboratories
West Point, Pennsylvania

Terry V. Iorns, Ph.D.
Iorns Consulting, Inc.
Mesa, Arizona

David Magnus, Ph.D.
Stanford University
Stanford, California

Tim Peakman, Ph.D.
U.K. Biobank
Manchester, United Kingdom

Eric D. Perakslis
Centocor (J&J)
Malvern, Pennsylvania

Jo Pisani
PricewaterhouseCoopers
London, United Kingdom

Sandy Weinberg, Ph.D.
Fast-Trak Vaccines
GE Healthcare
Upsala, Sweden

Acknowledgments

The editor extends his gratitude to all the authors, many of whom work actively in the industry, for their efforts and time to make this book possible.

Also, the editor is grateful to Michael Masiello and Anita Lekhwani of Taylor & Francis for their guidance and support during all stages of this book.

Contents

chapter 1

Drug Discovery in the Modern Age: How We Got Here and What Does It Mean?

Jeffrey S. Handen

Contents

1.1 *Introduction*

From Alexander Fleming's pioneering work with the discovery of penicillin to the no less monumental technical feat of mapping the human genome, the latter half of the 20th century held the promise of the development of drugs and other therapeutics for the final eradication of bacterial infectious diseases and the opening of the door to biotechnology-based medicines for the treatment and cure of disease at the very molecular level. Though the past 50 years have been marked by unheralded advances in the research and development of therapeutics, the expectations of a radical revolution in the treatment of disease have yet to come to fruition. The pharmaceutical industry, as a whole, is arguably suffering from an innovation deficit as it struggles to cope with new and ever-increasing economic pressures. For the period 1989 to 2000, only 15% of the 1035 new drug applications (NDAs) submitted to the Food and Drug Administration (FDA) were judged by the agency to be new molecular entities (NMEs) worthy of clinical priority review. The biotechnology industry has fared somewhat better, positioning itself as an engine of growth for the future. Currently, there are approximately 350 biotechnology-derived drugs on the market. That number is expected to increase by 50% within the next 5 years. However, it still remains to be seen if these drugs are better, faster, safer, and truly represent a new paradigm. What is clear, from numerous oft-cited statistics, is that drug discovery and development is getting longer, more expensive, and no better, suffering from the same clinical attrition and safety-related market-withdrawal rates today as it did 20 years ago. Moreover, the explosion of potential new targets from all the various *-omics* initiatives and endeavors, and the hunt for that elusive needle in a haystack, which now characterizes drug discovery and development, will only get more costly and more lengthy. For the full promises of the last 50 years to be realized, a new approach to drug discovery and development must be initiated. That approach can perhaps be best characterized as a modern-day back-to-the-basics approach—the intent being to understand the science, unravel the story, and then intelligently apply technology to bring to bear the weapons of industrialization, not just automation, in order to start with a bigger needle and a smaller haystack.

The aim of this book is to jointly explore such concepts, examine the current state of the art, look at some early success stories, and initiate a dialogue and challenge to the status quo. The goal is more specific targets, more selective compounds, and less time.

The modern pharmaceutical industry can trace its roots back to the chemical and dye manufacturing industry of the late 19th century, developed for the burgeoning textile industry of the industrial revolution. Early microscopists utilized these various chemicals and dyes to stain and enhance the semitransparent features of cells, protists, and bacteria. It was noted early on that some of these dyes showed marked selectivity and some showed marked bactericidal activity. Paul Ehrlick was the first to translate these two observations into the development of a chemotherapeutic agent, a chemical substance with affinity for a specific target. In this case, it was the development of a series of trivalent arsenical drugs active against spirochete-induced syphilis, ultimately resulting in the development and manufacture of "Neosalnarsan" for the treatment of human syphilis. Ehrlick's "magic bullet" concept and early championing of the concept of chemotherapy, along with his many other contributions in the field of immunology, led him to be awarded the Nobel Prize for Medicine and Physiology in 1908 and can arguably be cited as the start of the modern era of drug discovery and development. Ehrlich's laboratory synthesized and tested 913 compounds in the trivalent arsenic series before they came up with a compound with sufficient efficacy *and* manufacturability *and* delivery characteristics (1). Since that time, the numbers and shear volumes involved in the discovery process have increased orders of magnitudes, to today's corporate compound collections, which are measured in the millions.

The challenges faced nearly a century ago remain much the same today, namely, how to effectively and efficiently identify targets that are amenable to therapeutic intervention and how to design, test, and select candidate compounds that are not only safe and efficacious, but also have both favorable delivery (i.e., absorption, distribution, metabolism, elimination (ADME)) and manufacturing characteristics in a cost-effective manner.

A typical drug discovery and development program today approaches $900 million and 15 years (including the cost of failures), with *discovery* defined as target identification through lead optimization, accounting for $530 million and approximately 6.1 years of the total expenses (2). The remaining expenditures are accounted for during the preclinical and clinical development stages.

Drug discovery and the search for new molecular entities and novel active substances are still today arguably as much a serendipitous undertaking as one of rational design. Most organizations involved in drug discovery diversify their approaches and strategies to drug discovery in order to spread the risk involved.

Prior to the 20th century, medicines and pharmaceutical development were largely the domain of folk remedies and traditional healing practices, built on years of trial and error and observational, anecdotal knowledge passed down in oral traditions. The development of the smallpox vaccine by Edward Jenner in the late 18th century is an archetypal example of this folk medicine. Jenner noted that milkmaids who by their vocation were exposed to cowpox, a relatively benign bovine variant of smallpox, were conferred some type of resistance against the smallpox scourge sweeping Europe at the time. This, of course, led to the evocation of the concept of vaccination and subsequent development of the smallpox vaccine, and later development of the theory of immunization by Pasteur. This type of bioprospecting, as it has now come to be known, was perhaps the original source of drug discovery and one that is still in use today. For example, d-tubocurarine, a modified form of curare, a compound used by some Amazon Indian tribes as arrow poison, is used today during certain anesthetic procedures. However, though bioprospecting has provided many compounds for evaluation over the past century, and several companies have been formed solely to pursue this business model, it has been plagued by minimal success, even for the few promising natural product compounds identified. Unless a chemical synthesis can be developed, production is often plagued by complications in farming, and extracting consistent sources of compounds where metabolite composition and abundance can vary significantly in response to a wide host of environmental factors, such as time of harvest, precipitation, and age.

Screening is another strategy, and perhaps the most common one, generally employed in the pursuit of new compounds and active substances. In its simplest form, it entails the testing of a number of chemical entities for activity in some type of biological assay or screen. Initial primary screening usually involves the application of general screens designed to pick up broad classes of biological activity, such as receptor binding or calcium signaling. Active compounds are then subjected to rounds of secondary screening for confirmation and further elucidation of the biological activity. The advent of combinatorial chemistry and application of high-throughput technologies have revolutionized the speed and volumes with which companies routinely screen compounds. Today's corporate compound libraries are on the order of millions and can routinely be screened in weeks to months against a given assay.

General or random screening is the arbitrary testing of library collections, without any scientific rationale, to identify potential new candidate compounds in a hit-or-miss strategy. As compound collections increase, general screening has become more cost-prohibitive, and as knowledge of the chemical space has increased, general screening has given way mostly to the movement toward more focused or targeted screening. Targeted screening partially leverages the concept of rational drug design, utilizing an understanding of the structure-activity and structure-toxicity relationships of classes of compounds, along with mechanisms of interaction with potential targets, to restrict the universe of chemical space (or corporate compound collection) to a subset theorized to have a higher probability of exhibiting biological activity. The rationale for targeted screening can be derived from a host of sources, including the literature, proprietary structure-activity studies, and functional genomic studies.

True rational drug design, however, involves designing candidate compounds and three-dimensional structures to interact with a specific target, receptor, or biological pathway known to mediate a given pathology. The subsequent design is then tested. Rational drug design requires advanced understanding of the target, modeling, and simulation techniques. So-called in-silico drug design, while holding great promise, has to date yielded little success.

Serendipity is also an important source of drug discovery and can be a strategy. The accidental observations of beneficial unintended effects can and have been harvested to develop effective drugs. Sildenafil citrate (Viagra), a drug for erectile dysfunction, and Minoxidil (Rogaine), a drug for preventative hair loss, were both candidates in clinical development for other, ultimately failed, therapeutic interventions, in this case cardiovascular. The serendipitous observation of secondary biological, clinical activity led to these two drugs being marketed primarily for those indications. The influenza drug amantadine was accidentally observed to ameliorate symptoms of Parkinson's disease. Serendipity as a strategy can be implemented through effective observational science in laboratory and clinical studies, active surveillance of the medical literature and off-label use, and cross-disciplinary data analyses.

Traditionally, the pharmaceutical industry has discovered and developed small-molecule chemical entities as drugs. With the advent of biotechnology and the various *-omics*, the promise of large protein therapeutics is coming to fruition. Based on targets and compounds discovered in the human body, these biologics hold the promise of increased selectivity, increased potency, and thus better safety profiles as they mimic the body's natural physiology. Biologics thus are cheaper, safer, and faster to develop, as has been borne out. Biologics have a higher success rate in the clinics than small molecules (Figure 1.1) and generally faster times to development (Figure 1.2). In 2001, one third of all new entities approved in the U.S. were

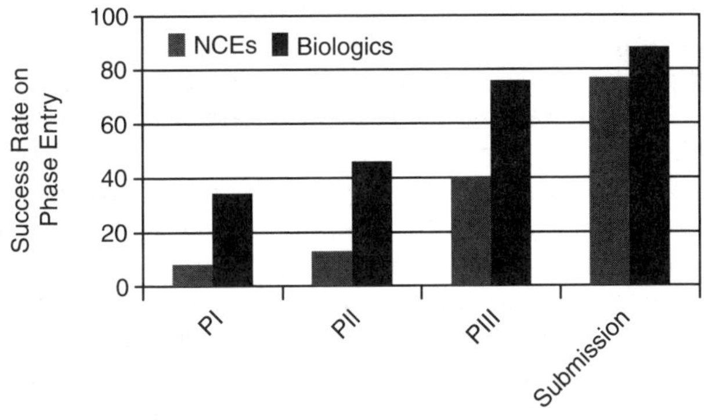

Figure 1.1 Cumulative commercialization success rates of small molecules vs. biologics. (From Industry Success Rates 2002, CMR International, Surrey, U.K., 2002.)

biologics; it is estimated that by 2005, 70% of all drugs on the market will be from biopharmaceuticals, and 20% of the top 100 drugs will be developed using biopharma research. Key to discovery of biological intervention is building the understanding of both the physiology and pathophysiology of a given disease state.

In reality, most modern pharmaceutical companies pursue, to varying extents, all of the discovery strategies discussed above. No single strategy has yielded significant increases in productivity to date. However, effective integration of these strategies with the total concept of industrialization and quality can effectively increase the probability of reaping the benefits of the more rational approaches, leveraging the burgeoning knowledge body of the *-omics*, to increase the size of the needle and decrease the size of the haystack.

The discovery stage of drug research and development, regardless of the particular strategy, has become increasingly characterized by the technological

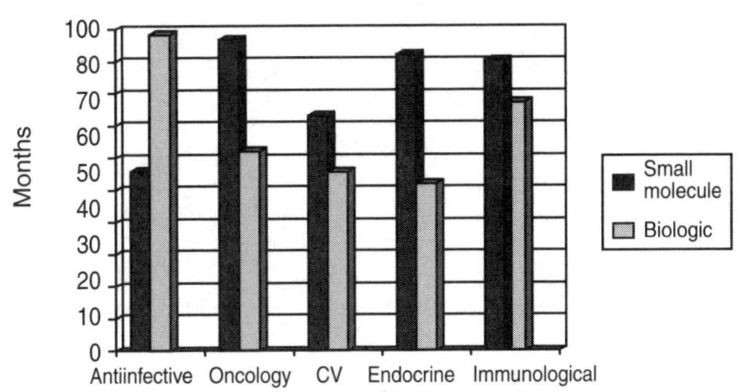

Figure 1.2 Mean time for clinical development for approved therapeutics, 1982 to 2001. (From J Reichert, Trends in development and approval times for new therapeutics in the United States, *Nat Rev Drug Discovery* 2:695–702, 2003.)

innovations of the past decades. High-throughput screening (HTS) has been progressing at unparalleled rates. Concomitant with this progress has been the explosion in genomic data, as a source of potential targets, and a corresponding explosion in the quantity of compounds available for testing, driven by advances in combinatorial chemistry. These factors are conspiring to pose both an unparalleled challenge and opportunity to the pharmaceutical industry. The opportunity is that for the first time in the history of modern drug discovery there are more potential targets and more potential leads than ever before. The challenge is that although HTS, combichem, and the fruits of the human genome mapping project are fueling this race for bigger drug development, the reality is that they have not yielded better drug development.

The R&D-driven pharmaceutical industry has not been able to successfully capitalize on and leverage the advances in technology of the past decades to ultimately commercialize more drugs, more efficiently. In 2002, 78 new drugs were approved for marketing in the U.S. by the FDA, continuing a steady downward trend. Just 17 of those new drug applications were characterized as NMEs, significant and innovative new drugs that contain an active substance never before approved in any form in the U.S. This number represents a 10-year low. This is in contrast to 152 approvals of supplemental NDAs for new or expanded uses of already approved drugs, a 67% increase from the previous year, and 321 approvals of generic equivalents (3). Indeed, spending on R&D has been increasing at approximately 10% a year for the last 20 years, but the number of new drugs launched each year has continued on a downward trend (Figure 1.3). This failure to translate technological advances into real gains in efficiencies has combined with various other pricing and cost-containment pressures facing the industry today to create an industry under fire—an industry that cannot meet the

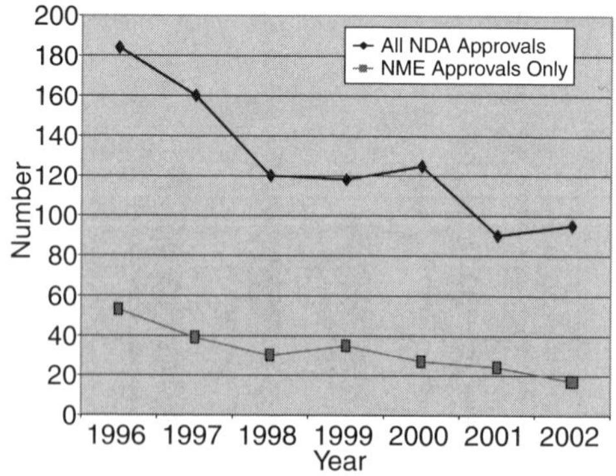

Figure 1.3 Number of new drug approvals per year, 1996 to 2002, U.S. (From Department of Health and Human Services, U.S. Food and Drug Administration, Center for Drug Evaluation and Research, 2002 Report to the Nation: Improving Public Health through Human Drugs, Department of Health and Human Services, Rockville, MD.)

historic demands of shareholder expectations for growth and an industry that is coming under increasing attack from payers, providers, and patients to justify often complex and seemingly disparate pricing policies on a global basis.

Indeed, for industry to maintain its historic double-digit growth and rates of return in the face of mounting R&D costs, companies will need to focus on bottom-line metrics to improve efficiency and throughput. Currently, the industry average for discovery output is five new lead molecules per year per 1000 discovery employees. The challenge will be to send 14 new lead molecules per year per 1000 employees. Currently, lead optimization to proof of concept takes 24 to 30 months. The challenge will be to shrink this time to 12 months. Presently, overall attrition rates stand at about 50% in preclinical development and 35% in late-stage clinical development. The challenge will be to reduce attrition rates to 15% in late-stage, high-cost clinical development. At present, approximately one third of R&D budgets are allocated to novel R&D efforts, with the rest actually focused on line extensions and "me too" drugs, with the accompanying lower-risk profile but also the accompanying lower rates of return. The challenge will be to allocate 50% of R&D budgets to novel discovery and development efforts in order to commercialize drugs that fulfill unmet medical needs, have novel mechanisms of actions, or represent significant advances in therapeutic intervention, all to command premium pricing. Shrinking time to market and improving the quality of the leads going into development will be critical to maintain shareholder value in the face of spiraling R&D costs (4).

Despite advances in technology, the cost of R&D continues to rise and far outstrip the cost of inflation. The reasons for this are many, complex, and controversial. Certainly, increased regulatory requirements and guidances have played a role, adding to the direct out-of-pocket cost of R&D, e.g., 21 CFR (Code of Federal Regulations) Part 11, requirements for designated special populations, and increased regulatory attention to QT correction (QTc) prolongation and cardiac risks associated with many noncardiac drugs. However, to blame the entire scope of increasing R&D on added regulatory burdens is an oversimplification and underestimation of the entire risk–benefit analysis of drug discovery and development from both a governmental and societal perspective and a commercial perspective. Safer drugs, though perhaps requiring greater up-front investments in parts of the value chain, not only benefit the patient, obviously, but also limit liability, increase compliance, and contribute overall to the bottom line. Indeed, there have been many regulatory initiatives over the past decade aimed at speeding up the drug discovery and development process and facilitating return on investment (ROI) for the research-based pharmaceutical industry. These have aimed at alleviating the regulatory burden and providing strategic avenues for exploitation, e.g., Prescription Drug User Fee Act (PDUFA), the pediatric exclusivity rule, and several of the Waxman–Hatch amendments. Yet drug commercialization clearly has neither decreased in complexity nor increased in efficiency. The average number of studies per NDA has increased from

30 in the early 1980s to 70 in the mid-1990s. The number of pages per NDA has increased from an average of 38,000 in the late 1970s to in excess of 100,000 in the mid-1990s. The average number of patients per NDA has increased from 1321 in the early 1980s to 4327 in the mid-1990s (6).

Perhaps of greater effect than the increased regulatory requirements are the continued high attrition rates at all stages of R&D, which necessitate high throughputs with all their associated costs. Significant investments in technology have not altered either the decision-making process or efficiencies, and have created a system where more assays are performed on the same compounds and targets without any gain in decision-making efficiencies. Most companies are quick to add the newest genomic and computational technologies to their arsenals of filters, screens, and models, yet rarely invest the necessary time and resources to fully understand the predictive capabilities of these new technologies and incorporate them into their project management and decision-making infrastructure. The net result is that new assays and technologies are continually added as part of the standard operating procedure, but old ones are rarely retired and they do not seem to be supplying any better information. Discovery output, as measured by investigatory new drug (IND) submissions to the FDA, has essentially been flat over the past 7 years (Figure 1.4). Likewise, the quality of the compounds moving into clinical development seems largely unchanged as the number of NDAs received by the FDA continues its downward trend and clinical attrition rates remain unchanged.

Confounding this lack of effective integration of new technologies to address productivity is the decreasing innovation seen in the pharmaceutical R&D industry. From 1989 to 2000 only 15% of the 1035 NDAs were judged by the FDA to be NMEs, worthy of clinical priority review (5). In the face of

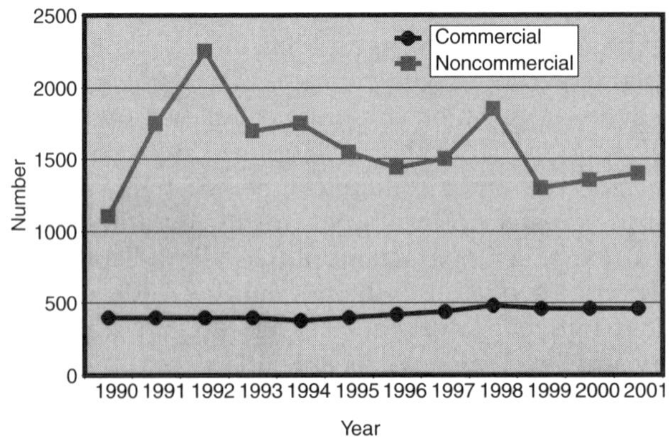

Figure 1.4 Number of original INDs received by the FDA each year. (From FDA/CDER, 2003 Report to the Nation: Improving Public Health through Human Drugs, Rockville, MD.)

increasing pressures and stressors, the industry finds itself caught in a conundrum. Targets and compounds, which are known to be developable, become the opportunities of choice, leading to the production of many me-too drugs and a dearth of truly innovative products. In actuality, this exacerbates the stressors on the industry because me-too drugs, though carrying lower risks, also lack market exclusivity and the ability to command premium pricing for innovative mechanisms of action and fulfilling unmet medical needs. One third of all R&D spending by the industry goes toward line extensions (6). This necessitates huge investments in sales and marketing efforts, in effect shifting the business model of the ethical drug manufacturer from a research-driven entity to a sales entity. Indeed, in addition to the nearly $65 billion drug companies spent on R&D in 2000 (7), the industry spent $13.2 billion on marketing to physicians just in the form of free samples and detailing, and an additional $2.4 billion on direct-to-consumer advertising (8). And pharmaceutical sales forces have increased exponentially to over 80,000 representatives in 2002 (9). Additionally, the sales productivity side of the industry, like the R&D side, has failed to gain any efficiency from economies of scale. Selling, general, and administrative (SGA) expenses—the costs associated with selling and the general expenses of running the business, reported on the income statement—are for most companies directly correlated to revenues (Figure 1.5). In other words, there is a linear relationship between the number of sales representatives and revenue. The only way companies have been able to improve revenue gain, on the sales side, like the R&D side, is to get bigger, leading to the continued Merger & Acquisition (M&A) activity characterizing the industry.

Thus, many companies find themselves mired in a vicious cycle where their lack of innovation, stemming from a failure to realize the efficiencies of their investments in R&D, commits them to supporting, in effect, two resource-intensive business models, instead of one: that of a research-driven ethical manufacturer committed to discovering and developing new drugs,

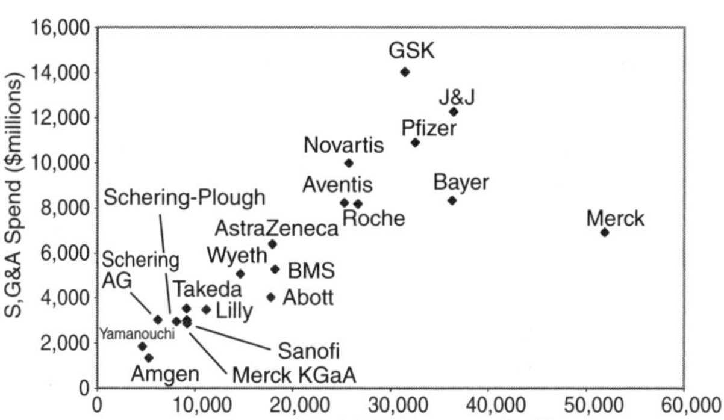

Figure 1.5 Revenue vs. SGA expenses for top pharmaceutical companies. (From Company Annual Reports.)

putting at risk hundreds of millions of dollars in the process, and that of a product development/commercialization company that likewise requires a similar investment to develop and maintain a massive sales and marketing infrastructure to compete in the me-too market, but against other nonresearch competitors (i.e., generics and other branded, noninnovative products) who do not have to bear the brunt of innovation.

For sales-driven organizations, like R&D-driven organizations, getting bigger has not resulted in getting better. This creates a huge and highly problematic behemoth to feed in order to maintain year-over-year end growth of 30% total shareholder return.

The question is then, Why, despite all the advances in robotics, miniaturization, computing power, and various other technologies, has the research pharmaceutical industry failed to gain ground on the bottom line: more, better, faster products? There is a growing realization that the available technology, in isolation, is not delivering the goods. Initial efforts to manage the vastly increased throughput of targets and discern the hits have been less than fruitful. They have been marked by difficulties in evaluating and implementing complex capital equipment and software within research-based user communities, problems in integrating new equipment piecemeal into existing facilities and departments, difficulties in working across functional/departmental/national/parochial boundaries, and ineffective attempts at using equipment as catalysts for cultural and organizational change. The cost of discovery continues to rise, and target time reductions are not being achieved.

Technological leaps and investments in discovery have proceeded at a dizzying rate. High-throughput screening infrastructures exist today that can generate 100,000 data points a day. Combined with an explosion of most corporate libraries of compound collection to the millions, and the burgeoning fruits of the various *-omics*, most discovery labs are screening more compounds than ever against more targets than ever and generating more data than ever. As recently as 3 years ago, the typical high-throughput screening lab was assaying on average about 15 targets (10), and all known drug targets to date comprised a universe of about 500 (11). Today, the universe of potential targets can be measured in the tens of thousands, though obviously not all of those represent true "druggable" targets. However, the magnitude of the scale illustrated in the high-throughput lab is just one example of the magnitude of the scale of the challenge. The product development life cycle is extremely long, on the order of years and decades, and extremely expensive, on the order of hundreds of millions of dollars. Applying technologies or reengineering as point solutions have to date resulted mostly in shifting the bottlenecks, which explains the M&A response of the industry, as economies of scale have been the solution applied to deal with these bottlenecks, opposed to truly creating a more efficient, holistic, integrated approach to the value chain. In addition, funding of technology investments has generally not taken into account the total costs of ownership. While R&D spending as a whole is approaching 20% of revenues for the pharmaceutical industry, R&D IT spending only accounts for only 7.5% of that 20% (12). Efficiencies gained from brute-force automation approaches have resulted in incremental efficiencies to the process. To truly revolutionize the drug discovery and development process, companies must address the entire value chain as an integrated undertaking and differentiate mere automation from true industrialization. Whereas investment in stand-alone robotics and technology serves to only automate current inefficient processes, industrialization provides the potential to truly reconfigure the landscape. Industrialization refers to a total quality approach with the emphasis not on throughput and quantity, but on quality and ROI. One study suggests that a holistic approach to data integration alone could save up to $149 million of the cost of developing a typical new drug (12). New systems also require enterprise-wide integration and communication, enabled by, but not defined by, technology. It has been estimated that up to 80% of the effort to install effective knowledge management systems, for instance, involved so-called change management: "adapting individual's work habits, thinking parameters, and communications skills rather than issues involving hardware or software." Additionally, studies of organizations that have successfully implemented enterprise-wide technology initiatives and reengineering efforts show that such organizations spent 10 to 15% of their project budget on change management resources (13).

In short, in a value chain as exceedingly complex as pharmaceutical drug discovery and development, any revolutionary increases in efficiency can only be accomplished by addressing the entire value chain en masse to build an efficient sustainable foundation enabled by appropriate technologies that *support*, not replace, efficient processes. Any point IT solution is a mere drop in the bucket and will fail in isolation to move the ultimate bottom line of $900 million and the 12 to 15 years it takes to get a drug to market.

What is clear, from numerous oft-cited statistics, is that drug discovery and development is getting longer, more expensive, and no better—suffering from the same clinical attrition and safety-related market-withdrawal rates today as it was 20 years ago. Moreover, the explosion of potential new targets from all the various *-omics* initiatives and endeavors, and the hunt for that elusive needle in a haystack, which now characterizes drug discovery and development, will only get more costly and more lengthy. For the full promises of the last 50 years to be realized, a new approach to drug discovery and development must be initiated. That approach can perhaps be best characterized as a modern-day back-to-the-basics approach. The intent is to understand the science, unravel the story, and then intelligently apply technology to bring to bear the weapons of industrialization, not just automation, in order to start with a bigger needle and a smaller haystack. The aim of this book is to jointly explore such concepts, to examine the current state of the art, to look at some early success stories, and to initiate a dialogue and challenge to the status quo. The goal is more specific targets, more selective compounds, and less time. The objective of this book is not to be a reference manual, but rather to instigate discussion and change through critical analysis. The chapters that follow jointly explore such concepts, examine the current state of the art, and look at some early success stories.

References

1. Nobel Lectures, Physiology or Medicine, 1901–1921. Amsterdam: Elsevier Publishing Company, 1901–1921.
2. P Guy, A Flanagan, J Altshuler, P Tollman, M Steiner. A Revolution in R&D, How Genomics and Genetics Are Transforming the Biopharmaceutical Industry. The Boston Consulting Group, November 26, 2001.
3. U.S. Department of Health and Human Services, Food and Drug Administration, Center for Drug Evaluation and Research. Report to the Nation: Improving Public Health through Human Drugs. Rockville, MD, 2002.
4. M Martorelli. Pharmaceutical Services Monitor: Ensuring Productivity Gains in Early Development. Investec, Inc., King of Prussia, PA, June 12, 2002.
5. The National Institute for Health Care Management Research and Education Foundation. Changing Patterns of Pharmaceutical Innovation. May 28, 2002.
6. Activities of the International Pharmaceutical Industry in 2000. CMR International, Surrey, U.K.
7. S Andrade. Using customer relationship management strategies. *Appl Clin Trials* 39–53, 2003.
8. The National Institute for Health Care Management Research and Education Foundation. Prescription Drugs and Mass Media Advertising. September 20, 2000.
9. L Sears. Sales reps, What's up doc? *PharmaVoice,* May/June 2002, pp. 46–53.
10. *Genetic Engineering News,* April 1, 2001.
11. J Drews. Genomic sciences and the medicine of tomorrow. *Nat Biotechnol* 14, 1516–1518, 1996.
12. Lion Biosciences/BCG Industry Study. Presented at Drug Discovery Technology. Boston, MA, 2002.
13. D Stiffler. How Consultants Changing Management Expertise Helped Ten Enterprise Software Implementations Succeed, AMR research report. AMR, May 2003.

chapter 2

The Regulatory Age

Sandy Weinberg

Contents

2.1 Introduction

In examining regulation, as in so many other arenas, it is all too easy to lose sight of the forest in the midst of so many trees. The number, import, and complexity of U.S. Food and Drug Administration (FDA) regulations, requirements, guidelines, and draft documents are often so overwhelming as to make it difficult to see the underlying principles that guide so many FDA actions.

There are three underlying precepts that together provide a foundation for and an understanding of all the myriad FDA actions: (1) proximal causality, (2) risk assessment, and (3) self-regulation. Those three principles can be used heuristically as well, providing a basis for prediction of future FDA actions and a projection of interpretations and emphases. Since the FDA's inception, these three principles have governed the ebb and flow of debate between proponents of the industry promotion and protection (providing more medical options) and the proponents of public health and safety (providing fewer but safer options).

The first of these trends is that of *proximal causality.* Consider what may be the most basic of FDA regulatory models: a consumer purchases a package clearly labeled as containing uncooked grains of rice. That consumer has a right to expect that the bag contains pure rice grains, free from insect parts, rat droppings, and other contaminants. Public health and consumer confidence are maintained as long as the label is accurate and the final product is clean.

But is it enough that the final rice package contains no rat feces or beetle parts? Does the consumer have a right to be certain that those unhealthy extraneous ingredients were not simply sifted out just prior to packing? Most consumers would respond affirmatively, and so regulators are authorized to investigate the packing plant itself to be certain that it is free of infestation.

And if that packing plant processes raw materials to eliminate those contaminants, should regulators also focus on the cleanliness and operation of that processing procedure? And on the equipment used in processing? And on the records of operation and maintenance of that equipment? And, perhaps, on the computer systems that store and interpret those records?

Much like the Russian dolls fitting within each other, regulatory attention focuses over time on the earlier, inner steps of preparation and process, always moving backward to more proximal causalities of safety threats and problems. In part, that movement is a function of satisfaction with the more surface levels of control; in part, it is a function of deeper understanding of the ultimate causes of problems. Arguably, an underlying causality of the trend derives from the reality that once investigators have achieved a satisfactory level of control for a surface issue, they have the time and resources to delve deeper: few regulators are likely to announce that the problem originally defined is now solved, and that their agency or position is no longer necessary. But most significantly, the trend to move backward to proximal causality is a response to consumer demand: standards of cleanliness are continually raised in response to public understanding. The same trend applies to the performance (and absence of side effects) of drugs, the reliability and therapeutic value of medical devices, the purity of biologics, and the quality and safety of all other FDA-regulated products and processes.

At some point, of course, the costs associated with growing demands reach a point at which cost becomes a limiting factor. Human blood provides an illustrative example. While most transfusion blood donors are unpaid, there is a cost associated with collection. As screening costs rise, the resulting expense involved in providing a qualifying pool of product rises proportionately. In a simplistic example, if it costs $1 to collect a bag of blood, and only 1 in 10 bags pass the screening criteria, the base cost per usable bag is $10. If those screening steps significantly increase the public health and safety, they are well worth any reasonable cost. But if the screening begins to provide only minimal value, then the value returned by the expense becomes more tenuous. Shall we exclude anyone who has visited the U.K. in the last 6 months, in response to the outbreak of bovine spongiform encephalitis, or mad cow disease? In the last year? Who has shared a meal with anyone who visited the U.K.? The incremental cost of each bag can begin to act as its own rationing factor, deriving the poor (or the poor selectors of insurance) from desirable treatment.

To avoid such a situation requires that, at some point, a risk assessment is used to weigh the value of further testing or screening against the relative safety gain (and increased cost) of that process. Regulators have always dealt with these equations: How many adverse reactions are acceptable for a drug that provides valued therapy? Should a vaccine that can save thousands in the (possible) event of a disease outbreak, but that causes severe life-threatening complications in a small number of cases, be widely administered? How does that equation change if the disease event is increasingly likely? If the complications are increasing severe, or increasingly common?

Risk assessment, the balancing of increased safety against the limited effect of the cost of implementing that safety factor, is the second major regulatory trend. Originating in the review of medical devices, the concept of risk (or hazard) analysis considers the probability and potential severity of an adverse patient reaction, system performance, or manufacturing process. Regulatory scrutiny is then assigned, and supporting evidence collected, in proportion to the degree of risk inherent in a given situation.

The risk assessment represents a rational triage of regulatory energies and an assignment of the expense involved in a major regulatory study to those areas of highest concern. The effect is a general lowering of noncritical regulation and a slight, albeit real, reduction in ultimate drug prices. As drug prices are increasingly criticized, and as the FDA's budget is stretched even thinner, this rational approach to apportioning regulatory efforts represents a significant trend.

The third major trend may surprise many members of the general public and even some long-term industry practioners: the FDA has long and consistently declared that it does not regulate in pharma industries. Rather, the FDA has taken the consistent position that the industry is *self-regulated*, and that the agency's job is to oversee and police this self-regulation.

All companies and products falling under the wide FDA umbrella are subject to internal quality assurance requirements. Even if a product performed flawlessly, without any side effects, purity defects, or adverse reactions, the manufacturer would be justly criticized if it did not have an appropriate quality assurance program in place. Anticipating the total quality management (TQM) and International Standards Organization (ISO) 9000 movements of the 1990s, the FDA and its industries have long established a goal of not just the equivalent of zero defects, but a quality system to continually improve and oversee the minimization of problems.

Working in conjunction with the proximity causality and risk assessment trends, this emphasis on self-regulation results in a continuing shift of attention backward in the laboratory and manufacturing processes. Users of Laboratory Information Management Systems (LIMS) now routinely investigate their software suppliers, since the internal quality of an automated laboratory is dependent increasingly on the accuracy of the software that manages and interprets experimental results. In much the same way, manufacturers extend their quality assurance to reach to vendors of raw materials, manufacturing equipment, and software control systems. The self-regulation, tempered by a risk assessment to put priority on areas of greatest potential danger, forces a proximal causality shift.

These three interlocking trends—proximal causality, risk assessment, and self-regulation—provide a framework for understanding current, evolving, and "far horizon" regulation. Together, they can lead to both an understanding and a logical prediction of the future of FDA actions in all areas, perhaps most dramatically as applied to the drug discovery and development process.

2.2 The Drug Discovery and Development Process

The FDA regulatory process applies to all aspects of the process of bringing drugs from laboratory to consumer—except, sometimes, the discovery process. The drug discovery and development process is much like a sieve, testing many compounds to select the few promising solutions that will enter the formal testing-to-production pipeline. The majority of those compounds, never patented and never pursued, are in practice exempt from all regulation. Only those substances that continue on to an FDA investigatory new drug (IND) and clinical testing and eventual manufacture and distribution are included in the regulatory process.

However, if a drug does continue on through the pipeline, there is a retrospective inclusion of its pre-IND and clinical development activities under the regulatory umbrella. In reality, then, only successful drugs are subject to regulation in the early stages, but developers are forced to prepare all prospective drugs for that eventual regulation. This retrospective regulation is another example of the self-regulated nature of the industry. A research organization anticipating producing a successful drug is wise to implement its own regulatory program from inception in order to prepare for the eventual review of regulatory documentation for the few surviving compounds.

The self-regulation of drug discovery and development, in anticipation of possible regulatory review should a drug be the subject of an eventual IND, is further complicated by a recently evolving development. A number of drugs are tested for a specific purpose and fail to meet the standards for further investment. Later, that rejected drug is found to have a secondary effect for which it does represent a viable product. For example, Viagra may have proven not to be the optimal solution to a search for a blood pressure reduction agent, but when reconsidered for a secondary quality-of-life condition, it emerged as a very successful product.

The reconsideration of rejected compounds again suggests self-regulation of the discovery and development process, to ensure supporting documentation for all compounds regardless of their immediate promise. In almost all situations the most successful and cost-effective strategy is to operate the drug discovery and development process in accordance with regulatory guidelines, anticipating the possibility of eventual retrospective regulation.

There is a significant exception. In the past, research universities have concentrated on basic research. In recent years, however, with the rise of technology transfer units in the major universities and the acceptance of entrepreneurship among science faculty, it has become not unusual for an academic research team to pursue the process through product development and licensure (and, in a few cases, formation of a company to bring the drug to market). The effect of incorporating the drug discovery and development process into the university setting is twofold.

First, traditionally and practically, university laboratories have been exempt from regulation. While this separation may have to be rethought in the future, there are no immediate plans for FDA review of academic research facilities, and few if any plans by universities to conform to the record keeping, training, and testing requirements of the FDA. In effect, the regulatory process for drugs developed in academic settings stops at the patent/licensure point.

Second, this separation places greater pressure on licensees—the drug companies—to themselves review not only the results of drug development, but also the controls and laboratory conditions that produced those results. University technology transfer offices are asked to produce Institutional Review Board records and, increasingly, laboratory equipment records, certifications, and installation qualification/operational qualification records. While the FDA is unlikely to look at academic laboratories, the self-regulation of the industry is rapidly moving down the line. A university interested in maximizing the value of its licenses would be advised to include appropriate regulatory evidence in its portfolio documentation. Similarly, small start-up companies whose strategy relies on acquisition upon proof of product should include self-implemented regulatory reviews in their operations.

The effect of this self-regulatory trend in drug discovery and development is to suggest that while compliance may on the surface seem optional, the practical reality is quite different. If you are operating an unsuccessful discovery program, regulation is unnecessary. But if your plans include the hope that the program will ultimately lead to the development of successful products, regulation, starting with self-regulation, is the only viable alternative.

2.3 *Historical Trends*

The history of the FDA is rich, varied, and interconnected with the major political reformist trends of the last century.* Buried in the detail of significant dates, regulatory revisions, and personalities are five threads that together weave the fabric of today's regulatory tapestry: (1) accuracy of labeling, (2) growth, (3) oppositional relationship, (4) pressure to speed the approval process, and (5) risk assessment.

From its initial emergence from political compromise and populist demand, the FDA has focused its attention on *accuracy of labeling*. The agency

* For an excellent descriptive history of the FDA, see Philip J. Hilts, *Protecting America's Health: The FDA, Business, and One Hundred Years of Regulation*, Knopf, New York, 2003.

has always been most concerned with making certain that consumers and health care professionals are provided with easy access to accurate information. The use of that information in making appropriate treatment and prevention decisions has always been a matter of much tension and debate. To what degree should the FDA restrict the choices a physician can make in determining the appropriate treatment for a patient? To what extent should a patient be allowed to make personal choices of diet, drug, or device? The stakeholders are in general agreement that the seller of a product should provide accurate information about the content, effect, and common side effects of a medication, and that a major role of the FDA is to ensure that this information is correct and accessible. But the screening and approval process, by necessity a restriction of that range of choices, is not a universally held construct. Most patients would like the widest range possible; most physicians trust their own ability to select from the widest possible range of treatments. Many manufacturers would prefer an open policy (within the range of common safety), in effect offering a variety of alternate treatments of differing effects and values (a few manufacturers realize that such an open market encourages competition and weakens market share, and would prefer a clear designation of best-practice products). Various consumer groups, with a range of trust in the judgment of the general population, the care competence and concern of physicians, and the honesty and greed of pharmaceutical companies, would recommend different approval and screening policies. But few, if any stakeholders, would argue that a manufacturer does not have a responsibility to provide accurate information, or that the FDA does not have as a prime responsibility the assurance of that accuracy.

The accuracy of labeling, then, has emerged as a significant FDA historical trend simply because of consensus. While stakeholders may debate other roles, and while the political will to enforce and fund those screening functions may fluctuate over time and administration, the agreement on the importance of ensuring accurate information is consistent and widespread.

The second historical thread defining regulation of the drug development process is *growth*. Whether there has been a general growth of "big government," or of the quality of regulations, is a matter of ongoing political debate. But the growth of the number and variety of products and companies subject to FDA regulation is a matter of record and accord. To the degree that the growth of the regulated has outpaced the growth of the regulatory agency (in both size and budget), a triage of regulation has emerged.

The discovery of new classes of pharmaceuticals, the development of biotechnology, and the invention of new medical devices have all swollen the U.S. pharmacopoeia and list of potential treatments. At the same time, increased public involvement in health care (for example, in the home diagnosis of illness, expanded from the oral thermometer of 20 years ago to a wide range of pregnancy, cholesterol, HIV, and other self-test kits), increased media attention focusing on emerging diseases (for example, severe acute respiratory syndrome [SARS]), and decreased tolerance for the normal risks of life have all created and enhanced regulatory pressures.

Increasing demand and all but static supply (of regulators and regulatory budgets) have resulted in increasingly long review times, decreasingly common field investigations and inspections, and a need to find ways of rationing regulatory energies without shortchanging public health and safety.

Yet another historical trend is difficult to characterize without using a loaded vocabulary. The relationship between the FDA and industry has traditionally been *oppositional*: Does that opposition represent obstructionism or independence? The answer probably lies in between the two extremes. In any case, most of the history of the FDA has been characterized by a skepticism concerning industry claims, with the FDA serving as a perceived barrier or strainer of new product approvals.

Such a relationship is not internationally universal, or even unique within the U.S. In Sweden, for example, the national regulatory FDA equivalent works with industry to help bring products to market (perhaps this cooperation is a result of the heavy investment of the Swedish government in national industry). In some other countries some pharma products are effectively nationalized (see the Netherlands' Vaccine Institute [RIVM] as a vaccine producer, for example). And within the U.S., the Federal Communications Commission (FCC) and Federal Aviation Administration (FAA) have been occasionally criticized for their roles in promoting or supporting the industries they regulate.

Few if any industry people would accuse the FDA of promoting the pharma industries. Interestingly, as the airline industry and energy industries have been adversely pinched by deregulation, there is a growing realization that the FDA, providing a level playing field by insisting on quality and evidence of effectiveness, may actually provide the established pharma companies with real services, both by discouraging competitive quality compromise and by serving as an effective (cost) barrier to emerging companies.*

The FDA's history includes a continued trend as a watchdog, standing in opposition to industry excesses and cut corners. That trend derives from the agency's origins in response to excesses in the meat-packing industry and remains firm today, wavering only slightly in response to contradictory consumer demands for access to new products and scandals about unsafe or ineffective treatments.

That dichotomy of popular (consumer fueling and fueled by media) demand for rapid release of promising and safe cures and treatments without regard to the science of testing and review represents another major historical trend in the Food and Drug Administration. Whenever a new disease receives significant media attention, the FDA is pressured to *speed the approval* process, even at the sacrifice of safety and effectiveness. In recent years, the tendency has been seen in HIV/AIDS treatments, terrorist response vaccines and treatments, Alzheimer's treatments, and SARS diagnostics.

* I have long argued that there are only two kinds of U.S. companies: those that are not regulated and wish they were, and those that are regulated and wish they were not.

The industry realizes (but the media ignores and the public misunderstands) that the FDA represents only a minor portion of the lengthy approval process. While there is no doubt room for increased efficiency and speed in review, halving the FDA's time to approval would represent only a minor reduction (perhaps a savings of 6 months to a year) in the 7- to 10-year discovery-to-market process. It is possible, of course, to place the blame for the entire 7- to 10-year duration on the FDA; because no safety testing, dosage studies, toxicity studies, and effectiveness control studies were required, one could randomly select a molecule or compound, package it, and bring it to market in weeks—only to see the potential patient harm, limited value, or side effects for years later. It is the range of quality and safety studies that take time, and while the FDA does have responsibility to review and analyze those studies, blaming the agency for the time required is the equivalent of a student blaming a dedicated and objective teacher for a low grade.

These four historical trends—a focus on accuracy of label, a trend toward growth of responsibility with limited assets, a tradition of opposition, and a pressure toward acceleration of approval—have produced the final trend, which is only recently receiving widespread attention. The first four trends force a rationing of regulatory effort and argue for focusing those limited efforts on areas of greatest quality concern to bring safe and effective products to market as rapidly as possible. The trend to use *risk assessment* as the rationing determinant is the result of those pressures.

In medical triage the likelihood of patient survival and recovery is utilized to rationalize limited treatment. Similarly, risk assessment utilizes the potential ultimate effect of adverse events (measured in both probability and severity) as a rationing factor in assigning regulatory efforts, and in allocating the resources necessary to provide supporting evidence in response to those regulatory efforts.

The review of medical devices has long relied on a risk assessment (also called a hazard analysis). Medical devices are classified based on an analysis of likely adverse effects (a malfunctioning x-ray unit producing excess exposure to radiation, for example) in terms of the probability (based on historical evidence and experience with predicate or related devices) and severity (the effect of that event on human health and safety). Those classifications then indicate the kinds of evidence required for approval (submission of a 510(k), formal premarket approval (PMA) clinical studies, and so forth).

In recent years, the FDA has recognized the logic of this approach and has expanded it to other regulatory areas. In 2003, for example, the FDA released guidelines calling for the use of a risk assessment in interpreting and conforming to the regulation 21 CFR (Code of Federal Regulations) Part 11, Electronic Records; Electronic Signatures. Arguably, the trend will continue, particularly under a commissioner who combines medical experience with expertise in economics, to other regulatory areas. If it is necessary to rationalize regulation in response to competing demands to improve speed to market without sacrifice of safety and quality, then risk assessment represents an effective and rational approach.

2.4 21 CFR Part 11

Part 11 of Section 21 of the Code of Federal Regulations is an evolutionary result of the historical trends affecting the Food and Drug Administration. Part 11 deals with electronic records; the systems that produce those records in automated laboratories, clinical testing environments, and manufacturing facilities; and the electronic approvals and signatures of those records. The regulation includes specifications for the validation and testing of systems, for archiving and retrieval of records, and for training and standard operating support of user personnel. In an automated world, Part 11 represents the current "capstone species" (pending continuing evolution) in the regulation of pharma products and companies.

In the proximal causality trend, Part 11 represents the regulation of a tool used in support of Good Manufacturing Practices (GMP), Good Laboratory Practices (GLP), and Good Clinical Practices (GCP) guidelines. In the risk assessment trend, Part 11 is clearly subject to a rationing of regulatory energy and evidence in accordance with the probability and severity of adverse events. In the mainstream of the self-regulation trend, Part 11 requires that an organization establish standards for quality and prove conformity to those standards. And in the historical evolution of regulation, Part 11 conforms with the initial focus on labeling (here, the accurate description of system requirements); the growth of regulation (in response to the growth of automation); the independence from and opposition to industry, imposing standards of quality and testing on the computer industry; attempts at speeding the time to market (spearheaded by automating the review process); and the use of risk assessment as a rationing factor, previously discussed in the melding of historical and regulatory trends.

In the drug discovery and development world, so strongly reliant on automated systems for analysis, testing, screening, and control, Part 11 represents not only the current capstone of regulatory and historical trends, but in many ways also the risk-defined, self-regulated definition of process and progress. Part 11 represents the risk-rationed definition of appropriate procedure and pathway.

2.5 Drug Discovery and Development Regulation

Since the drug discovery process involves a sieve sorting of promising molecules and compounds to identify potential practical therapeutics, the regulatory process is by necessity somewhat convoluted. For many drugs, abandoned before final stages of development, there is no regulation at all. But because it is not possible to predict accurately which drugs will survive the screening process, and because the costs of replicating early testing are significant, it is most practical to assume retrospective regulation of all developmental drugs. Since it is not possible to tell which pathways the FDA will eventually backtrack, it is is necessary to preserve and document all potential paths.

The emergent dominance of a risk assessment approach further complicates this "regulatory potential" approach. Consider a theoretical research organization that is investigating five possible drug compounds. Drugs 1, 3, and 5 may be dropped because of unusual toxicity, unacceptable side effects, or nontherapeutic potential. Drugs 2 and 4 may be pursued to an eventual IND submission. At the time of that regulatory impact documented evidence of compliance must be available for review.

But what depth of evidence is necessary? Before testing commences is it possible to predict the results of a future risk assessment, which will in turn indicate the extent and depth of regulatory concern, and the extent and depth of compliance evidence required?

In some cases, experience with similar compounds (predicates) may provide some useful indicators. In other cases, the very nature of the therapeutic problem or potential solution may indicate high risk. The complexities of dealing with and testing nonautologous human blood, for example, force high-risk consideration for all related products. But for many products, particularly those that are closest to the cutting edge of new research, these indicators may not be available.

The safe alternative is a regulatory default; that is, in absence of certainty, opt for a higher level of evidence of compliance. In effect, the retrospective nature of the regulatory situation is likely to force the highest-risk category of regulatory review on many or most drug development situations. Even though most pre-IND drugs will never be regulated at all, the prudent course is to assume not only future regulatory review for all, but also the highest risk level, and hence highest regulatory level, for all.

This strategy can be successful, but it is expensive. Treating all potential drugs as subject to the highest level of regulatory scrutiny negates the price-controlling triage of regulation and will greatly and unnecessarily inflate the costs of research and development.

There are some possible solutions. First, the risk assessment can be mitigated, particularly in the case of screening potential future drugs, with a balancing benefits analysis. Second, the regulation of the drug discovery process can focus increasingly on tools and procedures rather than screened compounds. And third, a more tightly defined link between the pure science of research centers and the business of drug development can move the regulatory process to later, more predicable stages.

2.6 *Benefits Analysis*

A risk (or hazard) analysis examines likely adverse scenarios and assesses the likelihood and severity of those scenarios. Missing from the assessment is a consideration of balancing benefits analysis.* Consider, for example, a system that is designed to consolidate and interpret blood processing test

* A benefit balance is an implied and important element in the medical device hazard analysis procedure.

results. Presumably, such a system would reject blood donation that exhibited evidence of HIV, hepatitis C, or other pathogens. But if a testing system reports trace levels of HIV—possibly a result of sample contamination, imprecise testing, or alternate sources of the enzyme used to indirectly measure the virus—should a blood donation be utilized? Prudence says no—that if even a trace contaminant is measured, even if that trace is likely an anomaly, the donation should be rejected. The severity is simply too high to be an acceptable risk.

If the testing equipment is validated, then how should the regulatory risk be assessed? The probability of a misread—of declaring that no meaningful levels of HIV are identifiable when the sample actually contains a minute and insignificant trace—may be low, though the severity of such an error may be high.

But what if the blood donation—likely safe, but with a low chance of viral contaminant—is critically necessary? What if the alternative to the treatment is a higher probability of severity?

2.7 *Risk Assessment*

Risk assessment is the inverse coin of benefits analysis. It examines the probability and severity of negative outcomes of a process, applicable to drug discovery, utilization of a pacemaker, manufacture of a drug, or any other process. In the regulatory world, however, a risk assessment has a unique role: it can help determine the amount of compliance evidence required for acceptance. In effect, a risk assessment of the drug discovery process (or any other process) can determine how closely the FDA will scrutinize, and how much time, energy, and expense should be dedicated to proving compliance.

To conduct a risk assessment, integrate these six steps into the process of planning and implementing a compliance strategy:

1. Utilizing the written requirement documentation for a specified system, device, computerized equipment, component, or application, identify the desired performance under each of the defined requirement applications.
2. Utilizing historical experiences with the specified system, device, computerized equipment, component, or application; or utilizing historical experiences related to predicate systems, devices, computerized equipment, components, or applications; or utilizing industry standards related to the specified system, device, computerized equipment, component, or application, determine the alternate undesired performances for each defined requirement application.
3. For both the desired performance and the undesired performances for each defined requirement, determine the probability of occurrence of each performance. Probability of occurrence can be calculated

utilizing historical experiences with the specified system, device, computerized equipment, component, or application; or utilizing historical experiences related to predicate systems, devices, computerized equipment, components, or applications; or utilizing industry standards related to the specified system, device, computerized equipment, component, or application.

4. Analyze each of the undesired performances to characterize the severity of that performance in terms of risk to human life, health, and safety. *High severity* is generally defined as loss of life, substantial loss of quality of life, or substantial disabling effect. *Medium severity* is generally defined as compromise of quality of life or some disabling effect. *Low severity* is generally defined as little or no effect on quality of life or on normal life activities.
5. For each undesired result calculate the risk according to the following table:

Severity	Probability	Risk
High	High	High
High	Medium	High
High	Low	Medium
Medium	High	Medium
Medium	Medium	Medium
Medium	Low	Low
Low	High	Medium
Low	Medium	Low
Low	Low	Low

6. Apply the table results to the validation protocol or policy to determine the appropriate level of testing and validation.

2.8 Risk Assessment and Regulation

In the process of regulating the drug discovery and development process the risk assessment provides a measure of degree of regulation. But to what does that regulatory scrutiny apply?

In the most basic sense, regulation is applied to all aspects of the development process, but two important caveats mitigate that reality. First, as described above, regulation of drug discovery and development is applied only retrospectively. That is, the end product and process are subject to FDA requirements only if the decision is made to move the discovered substance forward in the developmental process. Lines and paths that are abandoned or halted are not subject to regulatory review. Strategically, of course, that restriction may be of little practical value since all paths are considered potentially successful (or they would not have been initiated): most organizations therefore opt to integrate regulatory controls in anticipation of possible success.

The second caveat has a greater strategic and tactical value. While the FDA is interested in all aspects of drug discovery (with the degree and intensity as defined in the risk assessment), the practical focus for the past 10 years and the foreseeable future is upon the tools utilized in discovery. Just as an educator may try to judge a student's readiness to learn by assessing reading skills, regulatory attention focuses on the tools to assess the process. The student's reading skills and the tools are effective foci because they represent both fundamental elements in the process and historical indicators of potential problem.

Since the drug discovery and development process is largely based on the use of complex systems—automated laboratory equipment, molecular modeling software, statistical analysis systems, and the like—those tools tend to be computers. And it is the very nature of computers that causes the most regulatory concern.

Computer systems are complex, based on internalized and sometimes obscured decision rules. They are flexible, allowing multiple applications with appropriate changes in internal directions (programs). And, unlike more conventional paper trails, computers can overwrite the changes made to those decision rules, directions, and even databases, often leaving little or no markers indicating changes were made.

These characteristics, coupled with the industry's increased reliance on the use of computer systems, has led to a focusing of a significant portion of regulatory energy on the automated tools of the field. In the drug discovery arena, where reliance on computers is particularly high, the major regulatory focus, mitigated by risk assessment, has been on the systems that collect, analyze, manipulate, and report. The result of that focus is the newest of the FDA regulations, 21 CFR Part 11, "Electronic Records; Electronic Signatures."

Originally intended to define rules for accepting electronic signatures in lieu of human (paper) signatures on documents, Part 11 has been expanded to define virtually all aspects of computer control and the regulation of computer systems. The requirement includes standards for documentation of the validation (testing and managerial control) of systems; of the training and operating procedure assistance to be provided to users; and of security, change control, and disaster recovery assurances. Below is a checklist that outlines the broad scope of 21 CFR Part 11:

- System is used in support of drug-related research, laboratory analysis, clinical research, manufacturing, production, or tracking.
- Record or signature system has been subjected to an appropriate and thorough system validation audit.
- Audit conducted within 24 months.
- Audit conducted by independent or outside expert.

- Audit included review of:
  - Testing documentation
  - Development documentation
  - Standard operating procedure (SOP) documentation
- Change control.
- Archive.
- Disaster recovery.
- Use.
- Training.
- Audit trail review:
  - Archive
- Audit included inspection of operating environment.
- System validation documentation has been collected, including evidence of requirements and design approvals, testing, and implementation.
- Validation protocol.
- Validation team credentials.
- Development documentation:
  - Requirements/design document
  - Trace matrix
- Standard operating procedures:
  - Use
  - Training
  - Change control
  - Archive
  - Disaster recovery
  - Audit trail review
- Testing:
  - Border cases
  - Norm cases
  - Code review
- System inventory:
  - Hardware
  - Software
- Records are retained for appropriate length of time (generally 10 years or two generations beyond treatment duration) in machine-readable form.
- Records are retained for appropriate length of time (generally 10 years or two generations beyond treatment duration) in human-readable form.
- Records are retained in heatproof, fireproof, flood-protected environment, are appropriately labeled, and can be restored in reasonable (generally 72 hours) time.
- Procedures are in place to restrict access to data and records to appropriately authorized persons.

- Operations checks of system have been designed to ensure appropriate functioning of hardware and software.
- Audit trails (preferably electronic and protected; alternately, manual and carefully monitored) have been built into the system to detect and identify data changed, including tracking of time, date of change, change agent, and reason for authorized change.
- Electronic signatures are utilized only in systems with dual-level unique identifier authorizations.
- Password/password.
- Password/key.
- Password/biological.
- Other:
  - Electronic signatures are utilized only in systems with internal procedures to ensure that approved documents have not been modified (without authorization) from specified date and time.
  - Time system:
    - Zulu time
    - Greenwich Mean Time (GMT)
    - Location-affixed time
    - Single time zone
  - Date system:
    - International (dd/mm/yy)
    - U.S. (mm/dd/yy)
- A methodology has been implemented to ensure the validity of input data. Such methodologies might include dual confirmation of input, the use of check digits, internal norm confirmations, or other techniques.
- Systems users and administrators have received appropriate regulatory and functional training.
- System users and administrators have ready and constant access to appropriately comprehensive, clear, applicable, timely, and management-approved standard operating procedures.
- All aspects of the electronic records and electronic signature systems in place have been designed to provide a level of security and control equal to or exceeding the equivalent controls inherent to manual (paper) systems.

The Part 11 requirements provide guidance for control of the system tools used in the drug development process. In a very practical way, they serve as an outline of the FDA scrutiny of drug discovery and development. To the degree and depth defined by the risk assessment, the FDA is concerned with the aspects of drug discovery and development defined in Part 11. Evidence of compliance represents an insurance policy providing confidence that should a drug move along the pipeline to a new drug application (NDA), the Food and Drug Administration will accept supporting data from the early stages of the development process.

Below is a summary of the state-of-the-art regulatory expectation.

2.8.1 Drug Discovery Regulation Checklist

To ensure quality control and regulatory compliance, a drug discovery and development facility should document evidence of:

- *Risk assessment*: Determination of nondesired alternate outcomes and results of the drug discovery operation; calculation of the approximate probability of those occurrences, utilizing historical logs and results of predicate device operations; determination of the potential severity (in terms of direct threat to human health and safety) of those occurrences; and resulting categorization of the operation as low, medium, or high risk. If risk is determined to be low, other steps can be mitigated (limited testing in validation), eliminated (the vendor audit), or reduced (the performance qualification, or PQ).
- *Validation* of the operation, including:
  - *System validation* of all automated components, including documentation of system requirements and design; development of a trace matrix to ensure proper testing to those requirements and design elements; development and exercise of appropriate test scripts; a review of the code itself; and an analysis of the standard operating procedures in place to ensure appropriate use, archives, disaster recovery, change control, and training.
  - *Process validation*, focusing on possible contamination of media, equipment, and final product, including but not limited to bacterial contamination (both gram positive and gram negative), viral contamination, and material contamination.
- *Part 11 audit*: If the system is automated, the FDA regulation 21 CFR Part 11, "Electronic Records; Electronic Signatures," is applicable. Part 11 emphasizes:
  - An audit trail to track all data changes.
  - If electronic signatures are in use:
    - Dual confirmation of identity
    - Locking of document after signature
    - Unambiguous time/date stamp
- Archive (electronic and human readable) of all files:
  - Control of data accuracy
  - Appropriate training
- Audit of *system vendor*, either by an independent expert witness certifying compliance with appropriate regulations or by the end-user organization. Key elements in the audit are the criteria utilized and the credibility of the auditor (hence, the preference for independent

experts who have detailed knowledge of the vendor system and who have credibility for the regulatory agency).

- *Installation qualification* (IQ), using preestablished standards to ensure appropriate initial implementation of the system or systems. The IQ may be conducted by the vendor, end-user organization, or a combination of both.
- Initial and periodic *calibration*, in accordance with a metrology plan appropriate to the specific system or systems in use. Some systems are self-calibrating; a few others do not require recalibration after initial installation.
- *Operational qualification* (OQ) and *performance qualification* (PQ), sometimes performed separately and other times (in some circumstances and systems) combined. The OQ ensures that the system is ready for use; the PQ ensures that it is appropriately in use.
- *Problem report*: A system for reporting (and reviewing) any encountered malfunctions, necessary changes, or other problems.

2.9 Summary

In the current era of the regulatory age, focus on the drug discovery and development process is based on three general trends. First, there is a natural trend to move everything backward in regulatory concern to causal levels, resulting in a focus on the computer system tools used in drug discovery. That focus is most strongly reflected in 21 CFR Part 11, which defines the evidence of compliance necessary to utilize systems and accept the analysis and data that those systems provide.

Second, the intensity of the focus on systems (and on all other regulatory aspects) is tempered by a rational understanding that the significance of a danger should help define the degree of control of that danger. Cost controls suggest that investing in the collection of evidence or in expensive retesting should be mitigated by an analysis of the danger that an error represents. The result is the risk assessment, a tool used to determine the depth of control and compliance evidence necessary. If Part 11 tells what will be examined in an FDA visit, the risk assessment predicts the likelihood of that visit and the depth or duration of the scrutiny involved.

The final trend lies in the origins of the FDA and its self-defined (and appropriate) role. The agency does not directly regulate. Rather, the FDA defines its role as supervising the self-regulation of an industry whose economic and altruistic interests require high levels of quality control. The drug discovery process should be and is self-regulated: the FDA's role is to ensure the quality and consistency of that self-regulation.

So Part 11 directs the focus to the system tools that control the process of drug development and discovery. The risk assessment determines the intensity of that beam. And an understanding of the self-regulated nature of the industry defines the real regulators.

In this regulatory age the drug discovery process is appropriately regulated by drug discovery and development organizations, providing controls and evidence to a depth defined by a risk assessment, in accordance with standards defined in Part 11.

Reference

1. For an excellent descriptive history of the FDA, see Philip J. Hilts, *Protecting America's Health: The FDA, Business, and One Hundred Years of Regulation*, Knopf, New York, 2003.

chapter 3

Industrialization, Not Automation

Tim Peakman

Contents

3.1 Introduction

The productivity gap in the pipelines of the pharmaceutical industry is now well documented and accepted by industry executives, business analysts, and shareholders alike. This, despite the promised returns from large investments in the mid-1990s in high-throughput technologies for screening millions of compounds against the large numbers of new targets from genomics and later the human genome project. As the health care environment becomes increasingly difficult and organizations become increasingly large and complex, most are struggling to maintain their current level of productivity. However, recognizing the need for greater productivity, pharmaceutical executives are demanding significantly greater throughput from existing resources. Simply increasing the level of investment in technology, particularly process automation, will not deliver this. Success will be achieved through properly planned and implemented process industrialization in an innovative R&D-based environment. This chapter will highlight the key elements of industrialization that have not changed since the industrial revolution in England in the 18th century. It will discuss how industrialization has evolved and developed from the second industrial revolution, through Ford's production approaches to car manufacture, to modern manufacturing engineering disciplines now standard in many high-technology industries. Finally, it will address how modern industrial approaches can be applied to pharmaceutical discovery organizations to significantly impact the productivity gap.

3.2 Current Drivers: The Need for Change

It is now generally accepted by even the most optimistic of pharmaceutical executives that the historic levels of performance of their organizations are not being sustained. Despite year-on-year increases in the R&D budget (>$50 billion in 2003), the number of new active substances (NASs)—genuinely novel substances as opposed to line extensions, reformulations, or approvals for off-label indications—has declined significantly in recent years and is at best flat (36 in 2001). Numbers of NASs notwithstanding, the drive for innovation, reflected in high prioritization assignments by the Food and Drug Administration (FDA), is not filling the pipelines of major organizations, and consequently, overall approval times for new products are actually increasing (from a low of 12 months in 1998 to 17 months in 2001 (1)). To make matters worse, success rates for compounds entering the clinic have worsened considerably across all phases of drug discovery (1). Against an environment of cost constraint, patent expiry, and controlled prescribing by the purchasers of health care, there is no evidence of drug pipelines capable of maintaining the historic returns to shareholders. The evidence of this is already starting to be seen in the stock prices of the major organizations. Even in the general environment of depressed global stock markets of the late 1990s, the value of pharmaceutical stocks has declined in real terms over the last 18 months, with a consequent reduction in the returns to shareholders.

It is widely accepted by the industry that the productivity gap exists and has to be filled. Broadly, organizations are looking to fill the gap by selective in-licensing and alliances, by increasing investment in R&D, and ultimately by merger and acquisition. Specifically, the latter two are aimed at increasing scale and scope of therapeutic, scientific, or technological reach. However, as organizations get bigger and bigger, often the hoped for benefit of scale is not realized. In the worst case, the sheer size and geographical and therapeutic spread create fragmented organizations based around local sites. As with many strategies and plans in industry, the hoped for returns do not materialize.

One of the major areas for focus in the industry in the last 10 years has been significant R&D investment in technology to find either new gene targets in disease or new compounds as clinical candidates. Unfortunately, increasingly large and generally unplanned purchase and use of robotic technology has not had the hoped for impact in terms of overall drug discovery productivity. At best, the current situation has created a poor return on investment on the hardware and facilities. At worst, it has led to suggestions that high-throughput technology is not an answer to discovery productivity and encouraged pharmaceutical organizations to invest in the next technology fix. A good example of this situation is high-throughput screening (HTS) of gene targets against large compound collections using sophisticated robotics, but the principles and lessons apply equally to other high-throughput processes either currently operating or planned for the near future. These include high-throughput genomics, genotyping and single-nucleotide polymorphism (SNP) mapping, and preclinical compound profiling. There are many reasons for the current situation, but this chapter will focus on why the planning and implementation of automation for high-throughput technologies in discovery has failed to deliver the promised productivity gains into the clinical departments and how process industrialization can redress this gap.

High-throughput screening has led the automation era in the pharmaceutical industry. In the early to mid-1990s, organizations recognized the appeal of using automation to screen the hundreds of thousands of compounds in their compound collections against gene targets, rather than the tens to hundreds that could be screened manually. This, the argument went, would find the needle in the haystack, the novel molecule with a genuinely different structure that would provide the lead structure for the next clinical compound. Organizations invested big sums in increasingly larger automation platforms to store and screen their compounds. Over this period, HTS groups made huge strides in the ability to screen very high numbers of compounds in single screens and became extremely adept at high information content assay development (2) and miniaturization and automation (3). However, this improvement has been focused largely within the assay development and HTS departments and has not translated to a corresponding increase in overall discovery output.

The reason for this becomes clear when it is realized that the average utilization of the overall technology platform in HTS departments is between 2 and 7% of an installed capacity of some 1 million data points per day (4).

The explanation for the large discrepancy between installed HTS capacity and realized output is complex. It is unfair to blame the screening groups for not simply keeping their hardware platforms busier. In several cases we have examined, there was little backlog of work within the screening group, and the group's own performance metrics suggested that screens were run at a reasonable efficiency when a stable screen method and all the required components were available. Instead, the real issue with the performance of HTS groups currently lies in the broader process of target to lead. The HTS process is completely dependent upon multiple inputs, without any one of which no screening can begin. These inputs are assay development, protein production (or live cell production), compound management, and external deliveries from third-party vendors. Each of these inputs represents a separate project subject in many cases to the unpredictability that accompanies activities within discovery research. In addition, the lack of a production culture of quality assurance, quality control, process and element tolerance testing, process integration, scheduling and capacity planning, and approved and standardized handover criteria all add to the unpredictability and variability of the HTS process. Hence, HTS currently operates in a reactive sense: "When we have a protocol that is stable, and we have sufficient target protein, and we have the third party consumables and we have the appropriate HTS screening line—only then can we start the screen." Indeed, the low functional utilization rates that apply to HTS facilities are the manifestation of the response to very low predictability of the overall project timelines operating within the target-to-lead organization. In this environment, high throughput can only be achieved with a just-in-time capacity by maintaining a very high availability of HTS resource.

It is clear then that although the technology is capable of delivering the output stated and required, how it is implemented and used in a complex, largely unpredictable multicomponent process across global organizations is the real determinant of success. These lessons of industrialization have been learned in other high-technology industries and bear a striking degree of commonality. They all have their roots in the first and second industrial revolutions in England in the 18th century, and it is valuable to understand the main themes of industrialization from this context to understand the challenge and opportunity for pharmaceutical organizations as they try to address the productivity gap in their industry.

3.3 *The Birth of Industrialization: The First Industrial Revolution*

Industrialization as a term was first used to describe the massive and irreversible changes that took place in England in the second half of the 18th century. Many people associate the industrial revolution with technology: huge beam engines driven by steam, producing coal and iron that transformed England from an expansionist nation on the edge of Europe to a global empire. Technology was certainly a key enabler, but the industrial revolution, the process of industrializing Britain, was the result of a number of sometimes haphazard but completely dependent developments across the whole of a largely agrarian-based society. Ultimately, the first industrial revolution can be characterized as establishing the conditions necessary for the massive increases in productivity of existing products and the production of entirely new ones. As will be seen, these huge increases in productivity were not fully realized until the second industrial revolution. It is here that the parallel with the pharmaceutical industry will be seen. Huge increases in output to meet a market need are possible due to new technologies, but these will not be delivered until all the other elements of industrialization are addressed and implemented.

England in the early to mid-18th century had a number of small industries that were relatively productive at a local scale. These included soap and glass manufacture, brewing, and salt production. Perhaps chief among these was the manufacture and processing of wool that was carried out in dispersed regions of the country. This industry typified the others: local producers either processed the wool themselves in small amounts or provided it to local spinners and weavers. Occasionally, local organizations would be under the control of a wool master, but production was relatively small, and where products were transported, it was slow and costly. However, in the second half of the 18th century, events occurring across the country accelerated the pace of change to such an extent that historians date the start of the first industrial revolution to about 1760. The following changes were the four prerequisite elements of the first industrial revolution (5) (Figure 3.1).

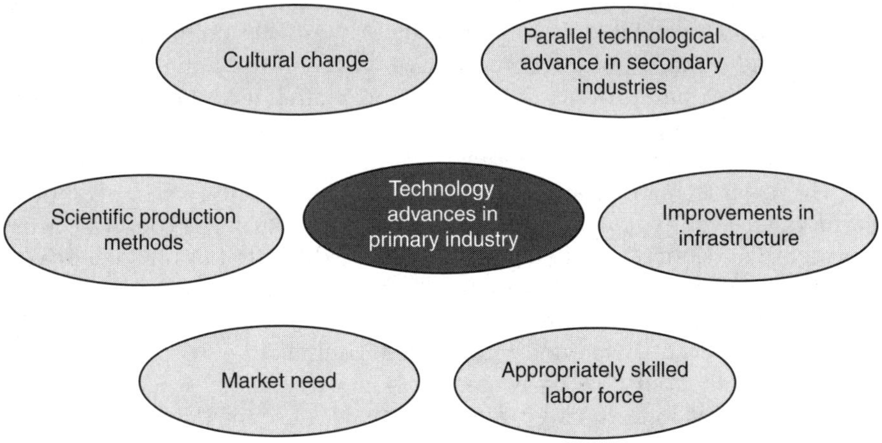

Figure 3.1 Changes necessary for industrialization. While technology advances in the primary industry are a key driver, the benefits of these cannot be achieved or maintained without changes in the other elements shown. This change is rarely smooth or entirely planned, but successful industrialization occurs when a rolling equilibrium is achieved.

Table 3.1 The Parallel and Interdependent Nature of Technology Development in the Cotton Industry

Year	Spinning	Weaving	Other Trades
1733		Kay's flying shuttle	
1740			Huntsman's steel casting
1767	Hargreaves' spinning jenny		
1769			
1779	Arkwright's water frame		
1782	Crompton's mule		
1783			Watt's rotary engine Cort and Onion—iron puddling
1785		Cartwright's power loom	

3.3.1 *Technological Advances in Primary and Parallel Industries*

It is indisputable that without the great technological advances of the period the industrial revolution could not have occurred in anything like the form it did. What is less well appreciated is that to do so required (1) the industrialization of associated supply industries, (2) an equilibrium to be established between different technologies in a given process to prevent bottlenecks occurring, and (3) entirely new technologies to meet needs created by the new processes (Table 3.1). Simply, there was little point being able to spin 10 times more wool if it could not be woven at the same rate. Equally, there was little point producing 10 times more cloth if it could not be treated, bleached, and dyed. The parallels here with high-throughput processes in modern drug discovery are obvious.

The major technological advances in the mid-18th century occurred not in the more conservative culture of the wool trade but in the newer cotton industry where, in 1733, John Kay's flying shuttle heralded the first technology-driven change on a well-established process. The flying shuttle could produce twice as much cloth per man as previous methods, but the supply of yarn had to keep up. However, it was not widely used until about 1760 due to human factors, ranging from lack of awareness of the technology to skepticism to outright hostility. In technology development, it is not the date of announcement that is important but the date that it is widely applied (8). Indeed, Kay encountered such hostility that he was forced to emigrate to France. Technology development and implementation is a characteristic of the first phase of industrialization. Many technologies, more or less bizarre, were invented that were never, if at all, adopted, which in part explains the relatively low levels of production compared to the second industrial revolution. It is only when mature technologies are fully implemented and integrated into a well-characterized process that high levels of productivity are achieved. After this first breakthrough, the technology enhancement and equilibrium gathered pace. Hargreaves' spinning jenny could simultaneously spin 80 threads on one loom. Arkwright's water frame was combined with the jenny by Crompton in 1779 into Crompton's mule to produce large amounts of very fine thread. At this point, the spinners were producing more thread than the weavers could use until Cartwright's power loom reestablished the equilibrium in 1785. The advances in cotton production needed new forms of power to drive them and to produce the machines. This required coal and iron, and following the production of Watt's rotary engines in 1782 (rotary motion being more suitable for the machinery), cotton production was transplanted to the coal-producing areas. This in turn drove further change as the old wooden machines were not strong enough for the new processes, which required iron frames and parts. Iron production using charcoal increased significantly until, when the charcoal ran out, Abraham Darby invented iron smelting using a coal-based process. The interdependence of parallel trades in iron production had started in 1740, when Huntsman developed a process for casting steel, and in 1783, Cort and Onion independently developed iron puddling, increasing production some 15-fold. The fact that this process was discovered simultaneously but independently was no accident. It was in response to a need. There were two other developments driven by the huge changes occurring in the various trades. Whereas preindustrial revolution production was distributed and carried out on a single or few machines, postindustrial revolution production was on a large scale overseen by owners of many machines colocated in single buildings. These were the first factories. Second, this period of tremendous industry and productivity improvement gave rise to equivalent growth in the bleaching, dyeing printing, and finishing trades. These organizations were the start of the modern chemical and ultimately modern pharmaceutical industries.

3.3.2 Improved Transport Infrastructure

One of the main reasons for preindustrial revolution manufacturing remaining largely local was the difficulty in transporting raw materials and finished goods. Goods could only be transported along rutted unmade roads or along the natural river waterways. Historically, many of the main centers of population had grown up on the major rivers, but these were not the focus of the new expansion in industry. Given this infrastructure, the quantities of raw materials required for industrialized production processes simply could not be sustained. Between 1760 and 1830, the quantity of raw cotton consumed increased from 8,000 to 100,000 tons, and iron output quadrupled to 1 million tons between 1800 and 1835. Therefore, to support this growth in traffic, the first industrial revolution saw a huge number of canals cut. Although these were still rather slow and transhipment of goods was difficult, they could be run directly between the major centers of supply of raw materials and manufacturing and linked with the main rivers to provide access to the seaports.

Despite their disadvantages, they carried the vast majority of raw materials in the first industrial revolution until about 1830, when they were superseded by railways. Almost coincident with the great canal era was that of road building. By 1830 there were 22,000 miles of turnpike roads built and operated by the local population. These groups contracted salaried engineers to build these roads, which saw the same development in technology as the major production industries. One of these engineers was John Loudon Macadam.

3.3.3 Population Growth and the Provision of Labor

Population growth in England in the first industrial revolution was both a requirement and an outcome of the changes described. The generally growing population created market demand for the products of the new production processes. Despite appalling levels of poverty in some sections of society, demand for the new products increased with the growing population that doubled in the 71 years from 1760 to 1831. This growing population supplied labor for the factories and parallel trades but, importantly, added another infrastructure requirement because so many people moved from the country to the growing towns. This had a number of impacts; of course they had to be housed near the factories, and the industrial towns of Northern and Midlands England grew faster at this period than at any time before or since. It can hardly be suggested that this movement was a model of town and community planning, but it still imposed major infrastructure requirements in these areas. Second, the role of the master craftsman was changing. Demand for handcrafted products made from start to finish by one individual was much less because of improved, standardized production methods and the workers in the factories became semiskilled operators in one or a few production steps. The conditions at the time also allowed unspeakable exploitation of unskilled workers, particularly children. Finally, the large numbers of people that moved from the land to the towns had to be fed. As will be seen, industrialization of farming took a long time to catch up to the productivity required.

3.3.4 Scientific Production Methods Including Distributed Manufacturing and Interchangeability of Parts

The final factor driving the first industrial revolution was a move to scientific production methods. As noted, prior to industrialization, production of noncommodity goods required a high skills base and was built from the system of trades apprenticeships. The methods used were nonscalable and resulted in low productivity levels. While two workshops would make broadly similar products, each would be made slightly differently, either entirely by hand or on jigs specific to the workshop. It was not, therefore, possible for an identical product to be produced and assembled from different workshops. Master craftsmen used (and rarely wrote down) their own methods, sourced all the raw material they needed, and coordinated and completed the entire process of production themselves. While there were many craftsmen who produced fine-quality goods, the majority were handmade items with small production runs that produced relatively expensive goods of variable quality. As the demand for greater output and quality grew with the industrial revolution, this changed. Production methods were characterized and standardized and tolerances were described; as a consequence, manufacturing became scalable. Plans and technical drawing specified the exact requirements of products. This was important because machines and parts produced in one part of the country had to fit machines in other parts of the country. They could not be fettled individually every time. This greater degree of standardization meant that manufacturing could be distributed to allow mass production of cheaper high-quality products. A good example of this is the development of firearms for the British Army. At the turn of the 17th century, the Baker flintlock was the first rifle to take advantage of barrel rifling, which gave a 10-fold increase in the effective range. However, because these weapons had to be handcrafted in Ezekiel Baker's factory, relatively small numbers were produced and, despite the known benefits, only a few regiments in the British Army were issued with them (9). By 1853 (strictly the second industrial revolution), however, the Enfield percussion cap rifle was the first precision mass-produced product in the world. Precision components were manufactured to high quality by machines in distributed centers by semiskilled men who specialized in one element of the finished gun (barrel, firing mechanism, stock, etc.). This process required coordination and scheduling, but the various pieces could then be brought together and assembled with much lower overall production costs and much higher productivity. Output, indeed, was so much higher that every regiment in the British Army was issued with the gun to such effect that army strategy was altered radically thereafter.

Even with the progress and change in these four elements, the industrialization of Britain would not have occurred without significant cultural and societal change. This is the fifth, and arguably the most important, aspect of the first phase of industrialization, without which the promised productivity gains would not be realized. Although the preceding sections describe significant change in a number of factors, it should not be inferred that this was a smooth process. Nor should it be inferred that suddenly England became an industrial society overnight. While it is true that those industries that adopted iron machinery and steam power saw big increases in their scale and output, of the four largest occupational groups at the end of this period (agriculture, building, domestic service, and shoemaking), none was highly mechanized. This period was a time of great unrest in England and Europe, which was exacerbated by the industrial revolution. The role of the master craftsman who understood and carried out the whole production process became diminished as the factories employed semiskilled laborers who completed one element of production to an acceptable quality and had no knowledge of the overall process. While some of the workers accepted the changes forced by industrialization and moved to work in the factories, others did not. Their resistance varied from a refusal to change to mechanized methods to violent resistance. This culminated in the activities of groups like the Luddites (actually several unconnected groups of tradesmen, but historically linked with the activities of Ned Lud) and the Chartists. The Luddites smashed machinery and burned factories to try to prevent the move toward mechanization and did so at the risk of hanging or transportation to colonies. The period also saw the rise of the scientist and engineer as highly skilled individuals who designed and planned the processes, machines, or great infrastructure and civil engineering projects, but did not actually carry out the work. As with any change, it is the human factors that were the most difficult to predict and manage.

This period highlights the prerequisites for industrialization of any industry: specifically, the establishment of conditions for increases in productivity (rather than actual massive increases in productivity) by the development, implementation, and wide-scale adoption of enabling technology in primary and parallel activities, development and alignment of infrastructure to support the primary and parallel activities, market demand and supply of the appropriately skilled workers, and trend to standardized scientific production methods. None of this would occur successfully without the cultural and human aspects of significant change being addressed at the same time.

3.4 The Second Industrial Revolution: A Period of Productivity

The first industrial revolution established the conditions for the massive increases in productivity of existing and new products during the second industrial revolution of 1830 to 1850. During this period, there were no new inventions with such a profound impact on production as those of the first industrial revolution. Mature technologies were established in controlled processes with supporting infrastructure with a suitably structured workforce, and current processes, technology, and infrastructure were improved or applied to new processes to increase throughput, quality, or speed. Chief among these improvements was the arrival of the railways, the planning and building of which exemplifies the effect of a rush to a technology seen as the next big wave of improvement. It was the building of railways that enabled the high-speed movement of goods and people across the country and the European continent. This in turn led to the spreading of ideas through journals, newspapers, and books and the standardization of machinery and time to enable a national rail system to function; this was the origination of Greenwich Mean Time, the point at the Greenwich Royal Observatory where time was set for the rest of the country.

The pace of railway building was tremendous: within 10 years, 5000 miles of railway had been built across the U.K. (8), but as with all industrialization, this was not without resistance. Originally, the period of railway construction was very expensive and carried mostly first-class passengers and light freight. However, as the volume of freight and second- and third-class transport increased dramatically, the road and canal owners increased their opposition significantly. By then though, it was too late; the era of rail had arrived. This is not to say that the period of railway building was a smoothly planned and implemented. In 1846, railway mania was at its height with its own party in Parliament. It was also the first national project that small investors put their own money into. During this period, more and more railways were built with little planning and no national coordination. Unlike France and Germany, there was no central planning, and for the first time, the technology equilibrium in industrialization had been artificially skewed. In 1847, the boom broke and led to a more considered approach with a planned infrastructure. The building of the railways was a huge logistical and supply project in its own right in terms of mobile labor, raw material, and general supplies. A further spin-off of the railway building program was the impetus it gave to agricultural reform. As the population grew and moved to the towns, the demands on agricultural production increased. During the period 1830 to 1850 agriculture was depressed and was struggling to feed the industrial towns. This was due to a number of factors, including the Corn Laws and poor land management and animal husbandry. As with other trades preindustrial revolution, there was little innovation or best practice. To address this, the government of the day repealed the Corn Laws, which stimulated reorganization. This in turn prompted specialization and a more scientific approach to farming, optimization of process, and an increase in diversified farming. Overall production and the area of land under productive agriculture increased significantly. The railways enabled the rapid transport of perishable goods to the towns and the transport of tools, chemicals, and new machinery back to the farms.

All of this drove the massive increases in industrial production. In the years from 1830 to 1850 consumption of raw cotton trebled to 300,000 tons, iron output doubled to 2 million tons, and coal production nearly trebled to 65 million tons. This was against a population growth of 50% in the corresponding 20 years (8).

A feature of the second industrial revolution was the changing shape of society. Although new trades employed much greater numbers of people (cotton workers were the third largest group of employees, woolmakers the seventh, colliers the ninth, and iron workers the seventeenth (8)), agriculture was still the leading employer. There were more blacksmiths than iron workers, and more men were employed on the roads with horses than on the railways. The reason for this is the main achievement of industrialization: the output per man per machine increased massively.

3.5 The Evolution of Industrialization: Frederick Winslow Taylor and Henry Ford

The drive to increase productivity continued into the 20th century and produced a number of individuals who started to take a more scientific approach to understanding the factors affecting levels of production and how they could be optimized. Perhaps the most famous (and arguably most misunderstood) is Frederick Winslow Taylor and his theory of scientific management. Taylor gave his seminal papers on scientific management theory in 1915 to the American Society of Engineers and to President Roosevelt's congressional hearings set up to examine how the nation's resources could be effectively used. Taylor developed his theories by observing how pig iron was loaded onto transport vehicles by gangs of men, and he quickly realized that the process could be quantitatively described and could be much improved by breaking it down to its constituent elements, optimizing them and then reintegrating them into a much improved overall process. These were, in effect, the first time and motion studies. He was able to increase the productivity of individual workers (with a concomitant increase in their wages) and the process very significantly. It is fair to say that the processes Taylor studied were manual, repeatable, and well defined. He also took a very hierarchical and, compared to modern views, none too enlightened view of the roles of the various workers. However, Taylor was the first to take an analytical approach that recognized that the process as a whole should be optimized, not just single elements, and that by specializing roles within the process, significant productivity gains could be made.

Perhaps the most common images of early 20th century industrialization are the production lines at the Ford Motor Company. Henry Ford was born in Michigan in 1863, and after early years in the Detroit Edison Company, he moved to the Detroit Automobile Company. At this time, cars were still built by hand, but Ford tried to introduce elements of industrialization: standardization of parts from gun manufacture and assembly line methods he had seen at Eastman's for photographic processing. In common with many pioneers of industrialization, he met with strong resistance. He left the Detroit Automobile Company and began to build his own racing cars. Success followed, and in 1903, he established the Ford Motor Company. Ford's goal had been to make cars available to the mass of the population—he described this as producing a car that his workers could afford. So, in 1908, the Model T Ford was produced. This was the first product of modern industrialization on such a scale—a simple, reliable car with no factory options (not even color), the design of which was unchanged from car number 1 to car number 15 million that rolled out of the factory in 1927. However, the early production runs produced cars that were still too expensive for his workers to afford, and Ford realized he would have to significantly increase the efficiency of his process in order to offer the car for a lower price. It was at this point that he really started to implement industrialization by designing interchangeable parts, optimizing the process, ensuring a continuous flow of process components, and specializing and dividing labor tasks.

Developing interchangeable parts meant that the production processes for these had to be significantly improved, but once the tools and jigs had been produced, a semiskilled worker could produce the parts to very high and repeatable quality standards. Ford optimized the process by removing any nonproductive time between actual assembly steps and, having seen the meat-packing houses of Chicago, established the layout of his factories so that the car parts were brought to the workers rather than them spending time fetching them. Then, by breaking the assembly process into 84 discrete parts, he trained his staff to one of these steps well. Finally, he commissioned Frederick Winslow Taylor to examine and optimize the *overall* process, and in 1913, the first Model Ts were produced on an assembly line. Ford's productivity increased enormously, and he was able to reduce the cost of his cars so that many of his workers did indeed buy their own cars. Even then, Ford still faced cultural resistance, being labeled by some a traitor to his class. Despite this, the first steps into modern industrialization had been irreversibly taken.

3.6 The Development of Modern Industrialization: The First Painful Steps

At the end of the Second World War, where industrial might was as important as military strength and industrialization had played a major part in supplying the munitions and hardware, people wanted consumer goods again. As tastes and technology moved on, so did the demands of the consumer: there is no better illustration of this in the context of modern industrialization than the car industry. Having a completely standard car with no choices was no longer acceptable; with the greater affluence of the 1950s, consumers wanted choices of shape, color, accessories, and so on, but of course they still had to be affordable. And so, learning the lessons of Ford but offering greater choice, modern car manufacturing developed and the major companies of the U.S. and postwar Europe rose to prominence. Raw materials were relatively cheap and competition was by and large confined to a national level. However, lead times were long, delivery time was highly variable, waste and rework were a major problem, and people were paid by their job titles rather than their skill and contribution levels. As consumerism continued into the late 1960s, American and European consumers were able to buy electronics from Japan that were high quality, reliable, and cheap. Then, Honda launched its range of scooters on the unexpecting U.S. and European markets, and suddenly Japanese manufacturing had a foothold and a reputation in previously unassailable markets. Finally, having laid the market groundwork, while U.S. and European car manufacturers were still producing large, often quite unreliable, vehicles, Japanese car manufacturers started to export high-quality products around the world that offered consumers choice at low prices. From a base of virtually zero market share outside of Japan, Japanese manufacturers now have no effective competition in certain sectors of the mass car market (6). Something in the way cars were being produced had changed dramatically that allowed manufacturers to offer choice and quality, ship the products halfway round the world, and still be able to undercut local producers on price for comparable products. This was the phenomena of Japanese industrial production methods: the story of attempts to reproduce this in the West is revealing and a lesson for the pharmaceutical industry.

Facing the challenge of dwindling markets, numerous trade delegations visited Japan to learn the secrets of its manufacturing capability. Executives were given tours around factories that, if not lights-out operations, were very highly automated. Pressed-steel body parts and components entered a line of robotic arms and a car appeared at the other end to be finally checked and finished by humans. The answer then was obvious: implement robotic processes. This would improve quality and reduce head count. Robots could be programmed to carry out multiple tasks and could be upgraded for new products or line extensions. Executives returned home and invested in robotics and established processes aimed at achieving high asset utilization. However, the expected productivity gains failed to materialize; in many cases, productivity declined. These early pioneering manufacturing organizations were the first to make the mistake of rushing to technology to solve a productivity issue without fully grasping the total requirements for success. They were first not because they were shortsighted, but because industrialization came to their industry early. Subsequently, many industries have repeated this mistake, and the pharmaceutical industry is no exception. It would seem that industries have to go through painful cycles of learning before they can claim to have a fully industrialized process that can be tangibly linked to an increase in productivity. This is usually driven by market conditions—in the case of car manufacture, the markets were being eroded in an environment of reducing margins. Productivity for the pharmaceutical industry was not traditionally an issue; margins were high, generic substitution was tolerable, and competition and price constraints were not so severe that a large industry could not prosper. Now, however, the situation is very different indeed. The current industry environment and likely future have been well described (1) and will not be discussed here. What is clear, though, is that the industry is facing huge pressures that are forcing it to address its productivity issue as a matter of urgency.

Many Western car and consumer goods manufacturing organizations are now world class and often superior to their Japanese competitors. This is because they have learned the lessons of industrialization and have either survived and prospered or succumbed to market forces or takeover. Visit any car showroom and you can specify exactly the car you want, from engine size, trim, accessories, and so forth, and a specific order will be initiated. This order will set in progress an entire supply chain of the specific parts required to build this car, and it will be delivered to you as specified to a high quality within 3 weeks. In effect, this hugely complex and interdependent process is delivering highly variable orders with minimal defects. How then, have these organizations evolved from the early efforts at industrialization to their current position? The answer is to address industrialization as a whole and not just technology—in other words, to address all those elements of industrialization described above:

- Technology development in primary and parallel industries.
- Infrastructure and process.
- Workforce and culture.
- Scientific production methods, including distributed manufacturing and interchangeable parts.

Addressing and developing these elements together has lead to modern manufacturing methodology.

3.7 Modern Manufacturing Methodology: Taguchi, Poke Yoke, and Statistical Process Control

Many organizations in the early phases of industrialization will rapidly purchase more and more expensive automation and try to fit it into existing processes and organization and expect productivity to increase. Even if productivity in the automated part does increase, the overall process productivity is rarely impacted. Now that high-throughput screening has become more mature in the pharmaceutical industry, new problems are being encountered around higher-throughput targets, supply of high-quality assays and reagents, and the capability to process hits coming out of the process. HTS has been embedded in discovery processes for long enough that we should expect to see an impact on the launch of new chemical entities. This is simply not the case. Piecemeal implementation of technology has long been recognized in more mature industries as a waste of resources. In those industries that have not appreciated this, often it is the technology that gets the blame for failure despite the fact it is capable of delivering the performance promised. This then has a pernicious effect on future rounds of budgeting, often resulting is a shift to the next technology "breakthrough."

If modern manufacturing methodology can be described briefly, it is about fully understanding the manufacturing system and its goals and how this will deliver the organization strategy before investment is committed. The manufacturing system refers to the process, the entire supply chain (not just the part within the scope of the particular production facility), the people, the organization, and supporting IT. Running through these elements is a focus on quality. Importantly, this is not quality for quality's sake, but rather quality that is fit for purpose.

3.8 Driving from the Top: Focus on Strategy

All modern manufacturing starts with a strategy: What is our strategy and how will we deliver it? No modern manufacturer would invest in automation and then try to implement it in the best way to fit the strategy.

Having set the strategy and the means of delivering, the organization will subdivide it to various parts of the organization. One of the key elements of industrialization is to separate R&D from operations. R&D is a vital exercise in many industries (not least the pharmaceutical industry) that is unpredictable, iterative, and innovation based. Such a process is rarely, if ever, suitable for industrialization. Manufacturing approaches are applied to those processes that are typically high throughput, reproducible, and standardized. This does not mean that they are inflexible; on the contrary, modern car manufacturers can produce up to 10,000 variants of a basic model. It is just that the components and processes that go into producing that model are highly characterized and understood. The systems are designed so that manpower, money, machines, and materials can be integrated, operated, and monitored to achieve

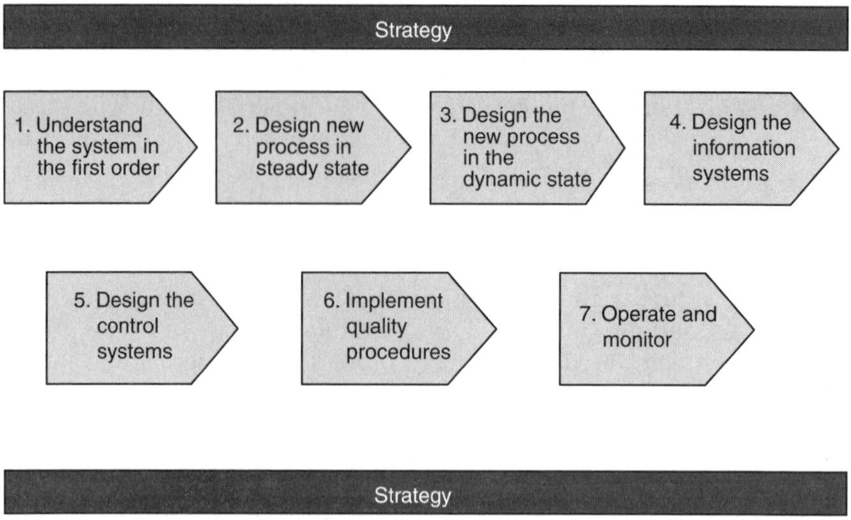

Figure 3.2 Overview of the steps involved in process industrialization. Note that no investment is made in hardware or software until the system interactions under real-world conditions have been determined and the capacity assessed against the strategy.

the overall objective efficiently. The first step to achieving this is to fully understand the system using a seven-step approach (7) (Figure 3.2).

3.8.1 Understand the System in First Order, Steady State

The first step is to describe the system and process in steady state at the first order—in other words, assume that no factors are limiting and that the flow of materials through the system is constant within defined limits of variation. While this process has its limitations, it describes the throughput of often complex processes in a logical series of steps with associated inputs, outputs, decision points, and information flows. The main problem with first-order approaches is that the interactions in dynamic, interlinked processes involving more than four or five discrete steps quickly become nonintuitive. Consequently, identifying bottlenecks before the process is designed becomes hard. Defining the system should aim for a level of detail that can be recognized by an expert in that system; otherwise, system description becomes an exercise in absurd detail. There are many methodologies for doing this (we have found that Integrated Computer Aided Manufacturing DEFinition (IDEF) works particularly well for discovery processes), but the method chosen should be used in conjunction with and validated by the people engaged in the existing processes.

3.8.2 Design the New Process in Steady State

The new process should then be designed on paper in the steady state, including material, manpower, and information flows and systems. It should be developed to deliver a certain capacity based on strategic resource management data that will identify throughputs and process times. It is critical to determine both process times and interprocess times (the times when a partially finished material is sitting waiting for the next stage of the process) to give total elapsed process time. Often, interprocess time can be as much as 50% of total process time. In highly automated processes, it can be up to 90%. Home personal computers ordered over the Internet are built and delivered within 3 to 5 days, of which 1 day is typically transport. Total build time is about 2 days and total process time (i.e., when specified components are assembled) is about 20 minutes. In the case of HTS, a typical primary-to-tertiary screen cycle is 3 months, of which 5% is actually in process time. It is natural to focus on optimizing process time; the real process efficiency can be achieved by examining and understanding the reasons for large interprocess times.

It is common, because of the functional structure of many discovery processes, that organizations will have good information on process times and total project elapsed times. Often they will have little or no information on interprocess times. In cases like this, it is still better for the people involved in the process to estimate interprocess times based on past experience, rather than ignore them or take a smoothed average based on elapsed time minus process times divided by the number of process steps.

This data-based approach will allow the system to be designed as a whole, identifying likely bottlenecks and potential solutions.

3.8.3 Design in the Dynamic State

Having framed the new process in the steady state, all manufacturing operations will use simulation to test it under real-world conditions. For continuous processes (as opposed to batch processes), the best method we have found is to use discrete event simulation (DES). This is a computer-based simulation approach that models the entire process and the interactions between the various elements within the process (human, technical, and organizational) and factors that will affect its performance. It relies on a good model of the system and its interactions; therefore, it has to be built on good data. There are two important elements about discrete event simulation: First, it is a linked system; in other words, a change in one part of the system will affect other parts of the system, which in turn will affect further parts of the system. This in turn will feed back onto the overall performance of the process. Second, it is based on constraints and business rules that prevent situations being measured that could not, in reality, occur.

Discrete event simulation can be used in two important ways: The first is to build a specific system capacity (a defined throughput at defined quality at defined cost), that is, to ask, "What total resources do we need to deliver a specific capacity?" It allows a number of combinations of resources to be assembled under different working patterns with real-world loadings to determine the best fit. Typically, resources are assembled and the overall throughput examined. The outputs of the model are graphically displayed so that the bottleneck is easily seen. The model then allows empirical de-bottlenecking of the system by shuffling system elements before any investment is committed. Often the results are surprising and nonintuitive. In a recent study of HTS facilities for a large pharmaceutical client, we observed that a specific piece of equipment was rate limiting for the client's assay portfolio. When a second platform was added, the bottleneck was removed but simply shifted elsewhere. By adjusting other elements in the system, the capacity was achieved. The advantage of this approach is that, providing it is based on good data, the results are a powerful adjunct to decision making. It also often results in a reduced budget request on hardware and facilities because the margin for error is greatly reduced. The organization in the case quoted saved approximately U.S. $3 million on a budget of U.S. $30 million. However, the funding was secured because of the power of the analysis.

The second use of DES is in scenario development to fully test the system to ensure it will deliver the installed capacity under any likely conditions. This is achieved by identifying all the variables within the system (e.g., in the case of HTS, these include, screen time, library size, number of targets, staffing levels, availability of technology platforms, and screening format) and then creating scenarios where these are varied. This allows the optimum design of the system as well as provides the ability to model the impact of new technologies. Here again, the answers are nonintuitive. There has been a trend toward miniaturization in screening. There is considerable debate about the value of this balancing cost of reagents against ease of miniaturization against quicker throughput. Using a detailed model of the whole system (rather than just the screening platforms), we showed there was minimal process advantage. While the time on the screening platforms was reduced, in the system we examined, the plate reformatting took up all of the time gained. Similarly, screening mixtures, as opposed to single compounds, had no overall impact on the process time, as any time gained in throughput on the screening platforms was lost in deconvolution of the hits.

The final important part of this stage is to design quality into the process—to build in quality assurance rather than rely solely on quality control. The quality standards should be fit for purpose. It does not necessarily follow that the optimum quality should be strived for; otherwise, the law of diminishing returns quickly applies. Building quality into the system will be discussed Section 3.9.

3.8.4 Design the Information Systems

Many pharmaceutical organizations who have implemented automation will have also invested in Laboratory Information Management Systems (LIMS) to manage the considerable data output from the screening campaigns. Many also have recognized the importance of an efficient inventory management and compound ordering and registration system. However, effective industrialization of the process requires several other information system (IS)

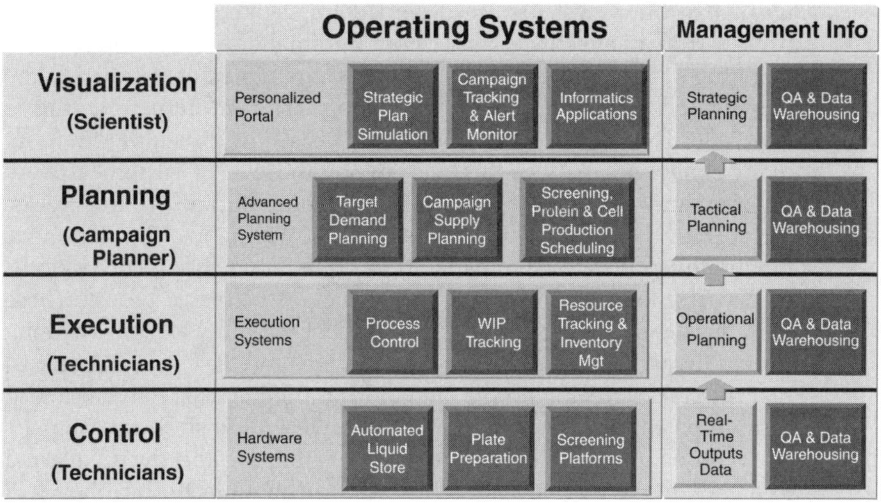

Figure 3.3 The multiple levels and functions of the software required to run an industrialized process.

elements to be planned and implemented through the process (Figure 3.3). Critically, organizations must also address what management information is required for the overall planning, scheduling, and control of the industrialized process. In an industrialized screening program, four levels of software need to be carefully integrated with the process and people involved:

Control. These are the discrete software packages that control the specific elements of the process. Typically, these will include software enabling the liquid compound store to retrieve and prepare compounds for screens, the plate formatting software, and the software controlling the individual screening platforms. Quality control data should be captured as part of these systems and all the data stored in a repository appropriate to the organization (Figure 3.3 shows a warehouse; federated approaches are also suitable). The management data are important for management of the current screening portfolio, but are also important for tactical and strategic planning based on real data.

Execution. These are systems that allow scientists to control and track the entire set of resources in the current screening portfolio. These will include systems monitoring the overall process and process quality, systems to identify work in progress (and hence identify problems when they occur), and sophisticated systems to track and measure the supply chain of *all* the elements required for an effective operation (i.e., not just the screens that are currently on the high-throughput platforms). In industrial operations, these systems are typically referred to as manufacturing execution systems (MES).

Planning. Execution systems are designed for delivering the day-to-day operation in an effective manner with minimal downtime and rework. Planning systems take a longer view and are aimed at optimizing the performance of the system over the long term. A forecast like this is truly effective if it covers 2 to 3 years. This seems difficult at first, but many businesses manage this by changing the level of the forecast over different horizons against which a frozen (short horizon of 2 to 4 weeks, high certainty, hard to change), slush (medium horizon of 1 to 4 months, moderate certainty, can be changed), or liquid (long horizon of 6 months to 2 years, low certainty, certain to change) plan can be set. These horizons are set based on the constraints of the process (i.e., how long it takes to carry out a single type of campaign) and the quality of forecast that the team agreeing on the demand can predict (e.g., how far ahead can we be sure of the exact target type we will wish to investigate). New campaigns are added to the system as they are approved. Referred to as advanced planning systems (APS) in manufacturing, they will build sophisticated models of capacity and demand based on business constraints (preventing unrealistic plans from being created). In doing so, they will plan the optimum use of resources by modeling the current availability and readiness of all the elements in a dynamic way. This is one of the key reasons why the planning level must be integrated with the execution level. As screens move through the process, or are delayed, or rescreen requests come in, the planning systems can reschedule based on availability of resources and the most up-to-date data. Once the new plan has been agreed upon, the system will schedule all the components necessary to deliver that plan against the business constraints of the entire system. For example, it may be possible to run a screen in 3 weeks' time because the screening platform is available. However, by having a good view of the protein inventory, it may be that repurification is required, which will take 6 weeks. In this case, the screen would not be scheduled. Similarly, a screening platform may be available but requires a 1536-plate format, which cannot be delivered because the formatter is already 100% utilized for the period. Clearly, without the connection to the execution and control layers, APS are useful only as long-term planning tools for agreeing targets (8,9).

Visualization. The final aspect of an integrated industrialized system is to provide the data in real time in an easily accessible and personalized form to those that need it. This will typically include a tool to view and modify the strategic plan developed by the APS level, project management tools to plan and alert staff involved in the process, and expert applications to allow manipulation of data as they are produced.

It is important that all four levels are addressed and planned. There is a tendency in pharmaceutical organizations to bespoke software. There is also a tendency to believe that the pharmaceutical discovery process is unique and no manufacturing-based software could possibly represent it. In the case of MES and APS, these are mistakes as robust off-the-shelf systems are available that have taken thousands of man-hours to develop for exactly the type of high-technology, unpredictable processes described.

3.8.5 Design the Control Systems

Once all other aspects of the new process have been addressed, the business rules, decision points, and quality control and quality assurance processes should be developed. The control systems have to be developed to give individuals at different levels information that will support decision making. Therefore, it is necessary to identify what information is needed by whom for which decisions. Data are needed for both project-related and planning decisions and quality and process control activities (see Section 3.9). Many organizations in manufacturing use a dashboard approach that displays the required data as an easily interpretable form for the specific user and allows drill down to greater detail, should it be required.

Once all of these elements have been addressed, a detailed implementation plan is required that will deliver a commissioned facility. This is not the same as the delivery and testing of the discrete elements in isolation. Commissioning refers to the proper integration and running of all the elements to deliver the required performance over steady state over a prolonged period. This is difficult enough given the complexity and multielement nature of the project, but to deliver this without disrupting the current business is indeed a challenge. It is unreasonable to expect a scientist to be able to manage this project; professional project managers working closely with a project team from all relevant disciplines are required.

3.9 Making It Work: The Total Quality Organization

If a successful manufacturing capacity is delivered, the focus shifts to one of continuous monitoring and improvement. This requires an emphasis on quality not just of products (or the outputs of the process), but also the intermediates, processes, and operations. Quality in all of these elements is required for the total quality organization. The total quality organization is delivered by focus on the following:

Organization. The agreement of simple, clear objectives for each individual and team, individual and team accountability, and relevant measures set off the overall objectives. Measurement of these goals is aimed at process improvement and not at sanction of individuals.

Methodology. Processes are developed, tested, and implemented, and the personnel properly trained. Depending on the environment, this can range from strict adherence to a standard operating procedure (SOP) to agreement on the deliverables.

Continuous process improvement. A focus from all personnel on continuous process improvement on the whole system. This is not, however, a license to "tinker" with existing processes. Process improvement needs to occur off-line (i.e., not interfere with existing processes), be tested for its results on the entire system, and, if positive, be properly integrated into the process.

This total quality organization was characterized following the work of W.E. Deming, J.M. Juran, and K. Ishikawa (indeed, Japanese manufacturing, unlike its Western counterparts, still has a Deming prize awarded annually). They identified that the organizations with the best quality levels focused on quality in the entire system (not just discrete elements), instituted in-process control and improvement (quality assurance [QA] as well as quality control [QC]), and avoided bureaucracy and empty exhortative slogans. Perhaps most characteristic of all was that these organizations involved all the staff in quality initiatives through multidisciplinary problem-solving teams. A lot has been written about this relatively simple approach, but one of the important outcomes is a shift to cellular manufacturing teams. Cellular manufacturing teams are groups of individuals who carry out a discrete part of the process, can treat any problem or task as a whole, and can adjust their resources or working patterns to meet particular needs: in other words, they are accountable for delivering agreed upon productivity. They need minimal coordination outside of the team, and consequently, day-to-day communication and decision making are easy. Cellular manufacturing teams can be characterized as follows:

- They are formed of natural process grouping, e.g., assay development, protein expression, automation specialists, screening specialists, and chemists.
- They require reduced levels of managerial staff above them.
- Members are expert, but no one is expert in all aspects.
- Performance measures are set for the team.
- Reward is based on performance and training/skills.
- The operators/team members are responsible and accountable for their output.
- They engage with internal and external suppliers.
- Operations and development balance long- and short-term goals.
- Permanent teams are formed for operational processes; temporary teams are formed for development and problem solving.
- They will use a variety of approaches to build quality into the process (see below).

Inculcating these elements into a pharmaceutical discovery organization is difficult, but the hardest element of all is developing the quality culture where quality concerns and productivity are part of the job. This strikes us as a paradox that a science-based environment based on measurement and validation of data and high standards of experimental methodology should have such a variable record on process quality.

3.10 Approaches to Implementing Process Quality

Process quality can be achieved in the design stage or during the actual running of the process. In the design stage, failure mode and effect analysis (FMEA) will try to identify likely failures, their cause, and the ultimate effect on the overall system (hence the need for simulation approaches as the system effects are rarely intuitive), e.g., high degree of pipetting variance for compound formatters. These possible failures can then be assessed as a weighted impact and can be eliminated or the process monitored carefully to ensure they are recognized early so that the overall system output is delivered to the required quality. Several approaches can then be used to eliminate or monitor failure. However, these approaches must be appropriately chosen: it is another common mistake of industrialization to be seduced by the latest management literature that full implementation of all approaches is required to achieve the required improvement. These approaches are summarized briefly here, but the interested reader should consult production management literature for greater depth:

Taguchi method. Often used after FMEA, this is an off-line approach using a variety of techniques focusing in more depth on specific elements of the process. In brief, it uses repeated rounds of experimentation to identify the optimum mix of variables in a system to produce a robust output (i.e., one that is not prone to small process variability resulting in large product variability). It is a valuable approach, but one that should be used in specific cases, often to focus on particularly critical elements identified in FMEA, e.g., high degree of variability in control activity in a cell-based assay.

Poke Yoke. If a responsive lean process is to be developed with minimal defects, QA approaches are vital. (QA prevents defects from occurring; QC detects them if they occur.) Poke Yoke uses foolproof devices or measures built into the process to detect defects. It shifts the emphasis from reducing defects to the customer to reducing defects *per se*. The measures or devices should be configured to detect any of three types of defects:

- *Contact defects.* Defects in the physical properties or configuration of components, e.g., assay elements.
- *Constant number defects.* To detect that a fixed number of movements or components has occurred in a process.
- *Performance sequence defects.* Ensures that all steps in a process have occurred or prevents incorrect steps from occurring.

Any defects are controlled either by halting the process (shutout type) or by raising an alert to the operator (attention type). These approaches, coupled with effective experimental QA procedures (such as Z' on assay plates), should ensure a much higher process and data quality.

Statistical process control (SPC). SPC is a method that ensures that process elements or products are delivered within tolerance, without having to measure every element. SPC relies on the simple observation that outputs of a process will be delivered with a natural variance. These variances fall within a normal distribution that can be determined by monitoring the process over time. If the tolerance of one element falls within this normal distribution, it is reasonable to say that all elements will fall within this tolerance and that the process is capable of achieving the required accuracy. Conversely, if they fall outside the distribution of tolerance, the overall product will be substandard. Industrialization recognizes that quality is not just a desirable for the customer, but costs a great deal of money if it is not built into the process. Measuring and controlling quality has become an extremely sophisticated discipline that has developed into approaches such as six sigma. Six sigma works by reducing the error rate of the process outputs by reducing the inherent variability. Large complex processes are particularly prone to high error rates because small errors can propagate through downstream tasks and produce high-output variability. Often, these errors cannot be detected by inspection of intermediates, so the source of the problem has to be identified quickly so that it can be resolved. Variability is described by the number of deviations from the mean value (sigmas) of the process intermediates and outputs. A one sigma deviation profile of the process includes 68.3% of the outputs, a two sigma deviation includes 95.5% of the outputs, and a six sigma profile includes 99.9997% of the outputs. A typical process has many points where errors and defects can be introduced. These points are referred to as opportunities. A three sigma process (99.7%) will have a defect rate of 3000 defects per million opportunities (DPMO). This means that every process step must have a defect rate far less than 3000 to produce an overall process with a DMPO of less than 3000. This sounds like a very high quality procedure, but if plating out a 1-million-compound screen set is considered, any small pipetting inaccuracies are going to have a significant impact on overall process quality. To our knowledge, no screening process has been modeled with the six sigma methodology. In industries where these measures have been developed, even a three or four sigma process will require 15 to 25% of total sales on non-value-add activities to sort out errors. At five sigma, the cost of quality drops to 5% of total sales. As with many other areas of industrialization of high-technology R&D processes, it is the telecommunications and computer industries that are leading the field. Since Motorola began quality approaches in 1981, it has eliminated 99.7% of the defects from its manufacturing processes. Perhaps the greatest quality levels are in the operations sector of airlines and airports. On the surface, a 99.9% perfect process sounds very good, but in fact it correlates to two plane crashes a day at Chicago's O'Hare airport.

3.11 Operating an Industrialized Process

Once a fully commissioned facility and process has been delivered, activity will switch from the project mode to a more conventional line mode of operating. If the industrialization has been properly configured, planning and scheduling will control the activities of the overall process. Inputs will be delivered on time in the right quantities at the right quality standards. Suppliers (internal and external) will have delivered appropriate raw materials or consumables in the right format to automation and people with the capacity to carry out the next steps in the process. When the process is complete, the data outputs will be fed to the scientists and will also inform the planning and scheduling and inventory systems. They will also be used to update portfolio and project plans, and process data will be used for quality assurance and quality control, and potentially for process redesign or error remediation.

Another common error is to try to play off suppliers on a purely price basis. Rather, using techniques such as Pareto analysis, key suppliers should be identified. These are typically providers of inputs that are vital to the process, high in value, perishable, or have long lead times. Here there is the opportunity to manage inventory, cost, and quality by engaging with the suppliers in long-term service-level agreements and including them in the planning approaches so that they have visibility of their production requirements. This will lead to quick response times, higher quality, and reduced cost.

The two other key elements of operating industrialized processes are technology upgrades and development and maintenance. Too often, particularly in the pharmaceutical industry, technology testing is done on the main screening platforms, disrupting the main efforts of the screening group. Additionally, these upgrades are often not properly tolerated and piloted to test the impact on the overall system. As we have seen, the impact can be hard to predict in isolation and unexpected bottlenecks may occur. New process technology should give a benefit either in scientific terms or in process terms. In either case, its impact and benefit should be modeled on the overall system. Some organizations will treat maintenance as a reactive process, calling out a service engineer when problems occur. This results in greater downtime, more expense, and much greater process unpredictability. Industrialized processes address maintenance preventatively by scheduling it as part of the overall master plan. Further, hardware is designed with a specific failure characteristic based on the simulation of the processes. Industrial automation providers now specify mean time between failures (MTBF) and mean time to repair (MTTR) as part of their product characteristics. These data are produced from significant factory and site acceptance tests.

3.12 Concluding Remarks

The preceding sections have addressed at some length how modern industrialization approaches are applied to deliver cost-effective, high-quality products. How this is achieved in high-throughput processes in the pharmaceutical industry has been mentioned and is discussed in much greater depth in reference 10 and Chapter 5 of this book. Key among these authors' conclusions are the cultural and human challenges of making this work in industries not used to these approaches. Here again, the parallels with the industrial revolution are striking.

Applying techniques from manufacturing in these environments will never be easy. However, the past years have seen these techniques applied in the service sector and to other high-technology R&D-based industries with measurable success. In these industries the arguments have been demonstrably proven with increases in productivity, quality, and speed of throughput. In summary, industrialization centers on having a clear technology strategy (which targets investment in technology platforms that support productivity without compromising innovation), where clearly defined pathways are used (without limiting the routes for creativity), where high-quality reagents and materials are used (to ensure global consistency of data outputs), and where common processes are followed (to generate more consistent results). It is still utterly dependent on good science producing validated druggable targets in relevant assay formats and producing high-quality lead candidates from the initial drug-like hits. However, if pharmaceutical discovery implements these approaches in an appropriate way, then the vision of a genuinely networked lead discovery organization is attainable and the productivity gains seen in other industries will be achieved.

References

1. SJ Arlington, S Hughes, J Palo, E Shui. Pharma 2010: the future of the pharmaceutical industry. IBM. NY, 2002, pp. 1–15.
2. M Beggs, AC Long. High throughpout genomics and drug discovery: parallel universes or a continuum? *Drug Discov World* 3:75–80, 2002.
3. M Beggs. HTS: where next? *Drug Discov World* 2:25–30, 2000.
4. M Beggs. Is Science Unmanageable? Paper presented at R&D Leaders Forum, Geneva, 2002.
5. A Briggs. *A Social History of England*, 2nd ed. London: Penguin, 1987.
6. A Webb. *Project Management for Successful Product Innovation*, 1st ed. Aldershot, U.K.: Gower Publishing Ltd., 1994, p. 346.
7. *The Lucas Manufacturing Systems Engineering Handbook 1998*. Internal handbook of the Lucas engineering organization.
8. S Franks. Business modelling and collaborative planning: the key to ever increasing productivity in the new millenium, part one. *Eur Pharm Rev* 67–72, Spring 1999.
9. S Franks. Business modelling and collaborative planning: the key to ever increasing productivity in the new millenium, part two. *Eur Pharm Rev* 70–75, Spring 1999.
10. T Peakman, S Franks, C White, M Beggs. Delivering the power of discovery in large pharmaceutical organisations. *Drug Discov Today* 8:203–211, 2003.

chapter 4

Compound Management

Eric D. Perakslis and Terry V. Iorns

Contents

4.1 *History and Evolution of Compound Management*

In the past decade, as high-throughput screening has developed as an accepted drug discovery activity, compound management has evolved from a simple dispensary function to a strategic activity (1). The recognition that the compound collection represents the "crown jewels" of the organization has led to the need to invest considerable talent and resources in managing and utilizing this asset. Indeed, with all groups within the companies dependent on the compound management activities and the increasing recognition that the makeup of the compound collection is of strategic importance, investment in compound management activities is increasing dramatically.

4.1.1 *Definition of Compound Management*

Compound or substance management is a relatively new discipline within drug discovery and agrochemical research organizations that deals with the acquisition, management, storage, processing, and distribution of research samples. Depending on the size of a research sample collection, the compound management organization will contain a set of systems and processes specifically designed to meet the substance evaluation goals of all areas of research.

4.1.2 *Why Compound Management?*

The evolution of the compound management function into a separate and distinct discipline in the drug discovery value chain is directly tied to the introduction and growth of high-throughput biological screening technologies. The transition was forced by the need of high-throughput screening (HTS) organizations for large numbers of compounds from many different sources. This contrasts with the previous close relationships between individual biologists and chemists in earlier organizations. This change in need has resulted in a compound management organization being formed in most drug discovery organizations.

The changing needs can be simply summarized with *better, faster, cheaper* access to compounds. The discovery organization, both HTS and therapeutic area biologists, needs better access to large numbers of compounds from various sources, including internal collection, newly synthesized, and purchased, and compounds obtained from various partnerships (more on various sources later). The race to shorter development times is a major driver for the industry, and faster access to compounds can help the discovery organization to shorten its timelines. Finally, with increasing cost pressure and tightening budgets, improved compound management represents an initiative where costs can be decreased while improving overall performance.

Prior to the ability to generate and screen huge numbers of molecules, the compound collection of any large company was typically in the ranges of tens of thousands of substances. These numbers were manageable in a typical pharmacy-like setup or stockroom. Inventories were tracked on paper originally, and later, as the collections grew, in increasingly sophisticated computer systems. Advances in parallel chemical synthesis and other technologies, coupled with the increase in throughput of screening capacities, spurred the growth of drug discovery collections and the ability to screen hundreds of thousands to millions of compounds. These numbers escalated the needs of these simple processes of storing and retrieval of substances to truly industrial-scale capacities.

In addition to the changing scale, the variety of sample formats also has and continues to evolve. Miniaturization and nanotechnology have facilitated the adoption of smaller, denser sample formats. Originally, samples were delivered individually in small vials and screening was carried out in test tubes. Automation of screening processes has led to microtiter plate-based screening formats that have evolved from 96 samples per plate delivered in microliter quantities all the way to plates containing 1536 (and greater) samples per plate in the same footprint, delivered in nanoliter quantities. A total of 384 wells per plate can now be considered the industry standard.

The variety of sample formats increases further when factors such as varying concentrations, solvents, sample types (small molecules vs. natural products), screening strategies (sample pooling/multiplexing), and permutations of all the above are figured into the equation. Additional complexity arises when different parts of the research organization require compounds in different formats and a single compound logistics area is serving the needs of many therapeutic teams and screening groups working on a wide range of biological target and assay types.

Another reason for a broad compound management organization is the ability to provide cost-effective services to accelerate the speed of drug discovery. Services provided by a compound management function represent tasks that research scientists do not need to perform, allowing them to focus their efforts in areas where only they have the scientific training. Thus, help in distributing new compounds to biologists frees chemists from this mundane task. Help in making compound solutions and hit-picking for confirmation and dose–response studies frees the biologists for the activity measurements and additional screens.

4.1.3 Use of Solubilized Compounds

The most common use of compounds solubilized in dimethyl sulfoxide (DMSO) is for screening. High-throughput screening relies on large collections of compounds to identify hits that display interesting activity in an assay. After the hits are identified, most organizations have a regimented process they follow to determine which hits are of most interest. This qualification process generally starts with a confirmation test. This is generally a repeat of the assay with a group of hit compounds to determine if the activity can be repeated. Hits that do not confirm are generally dropped from further consideration.

A variation to this screen approach used at some companies is the use of multiplexed samples or pooling. In multiplexed screening, each well contains multiple compounds. Any number of compounds can be used, but five is a common number. If a hit is observed on a well, there is an additional step required, deconvolution, to determine which compound in the well is causing the activity. Multiplexed screening assumes that the activity of the mixture is from the individual activity of the compounds and that no compound–compound interactions occur. Also, pooling usually requires that each compound be present in smaller concentration, resulting in decreased sensitivity of the assay. It is best if the compounds in a well are not structurally similar to avoid the additional complication of two compounds in a well independently showing activity.

By having multiple compounds in a well, multiplexed screening has the advantage that the number of wells to screen can be decreased. This has important time-of-initial-screening and cost-of-reagent benefits, particularly with some assays. The disadvantage is the need for an extra step, commonly called deconvolution, that is not easy and can end up taking a lot of time. Deconvolution usually involves separately running each compound from the multiplexed well producing a hit to determine which individual compound is responsible for the activity. Yet another alternative is multiplexed screening with orthogonal pooling. In this approach, each compound in the mixture is present in two wells, and the second well does not contain any of the compounds from the first well. In this approach, hits will come in pairs of wells and the compound causing the activity can be identified by a computer algorithm without the need for manual deconvolution. This approach, however, sacrifices one half of the benefits because each compound must be present in two wells. Both approaches to multiplexing or pooling require more sophistication in replication systems and in computer data systems. This can make the cost of implementing multiplexing prohibitive if existing systems are incompatible with the concept.

The next step is often to get a more detailed understanding of the activity profile by performing a dose–response study in which the same compound is tested in the primary assay at several different concentrations.

At this point, the process usually divides down two different pathways. Chemists review the data thus far, trying to build an understanding of how the chemical structure can influence activity. Chemists try to prepare additional compounds in a rational manner to increase the activity while optimizing other characteristics of the compounds to make a good drug. Biologists continue evaluating the hits they have using secondary assays, profiling assays, and sometimes initial animal studies. Then, as the chemists prepare new compounds, they are subjected to the same testing cycle, gathering more and more data to build a better understanding of drug activity.

This research by the chemists and biologists often leads to a frenzy of activity in the compound management organization. The chemists synthesize and submit many new compounds, and the biologists or screeners request custom plates of these compounds to evaluate in the primary assay. Those compounds with interesting activity are requested again for confirmation. The biologists request dose–response custom plates for the confirmed hits, and then they request additional custom plates or custom solutions for their secondary assays or preliminary animal studies. At some point, the biologists will also request neat samples of the most interesting compounds for additional follow-up studies. Coincident with this, other research groups want to evaluate all the new compounds in their screens, so the compounds must be plated into libraries. Multiply this activity across all therapeutic areas and sites of a large pharmaceutical company, and the amount of work can be overwhelming.

In most compound management organizations, most of this frenetic activity can be broken down into two major processes—library management and custom plates—and managed in a routine manner.

4.1.4 Compound Collection Valuation

Regardless of the models used, any large compound collection represents a significant investment and asset for an organization. There are several commonsense rules that can be applied to this valuation. The primary cost factors are novelty, purity and amount, and ownership.

Novelty refers to how unusual the chemical structure is or how difficult the compound is to synthesize. The more novel the compound, the more costly it is to obtain.

As one would expect, as purity and amount increase, so does the cost. At many companies, the emphasis has been on the number of compounds in the collection, and parallel synthesis and purification techniques are used to quickly increase the number of compounds in the collection, often with decreased requirements for amount and purity. Purity is not too important in HTS as long as the activity is due to compound and not some impurity; however, pure compounds are preferred for follow-up testing. Other companies emphasize the purity and amount of each compound in the collection, and this increases the costs. Factors include increased chemist time due to additional purification steps and less automation, higher reagent prices for higher reagent purity, more reagent, advanced synthesis and purification methods development, and larger and more complex equipment setups. It is difficult to estimate the additional cost to synthesize 50 vs. 5 mg of compound. For a difficult reaction, the cost may be several times, whereas for a scalable reaction, only a small additional cost of reagents is incurred. Five milligrams of a compound is usually enough to get it into HTS, but little compound will be available for follow-up testing. A more satisfactory goal is to synthesize 25 to 50 mg of compound, providing enough compound for more than one solubilization, plus providing adequate compound for various follow-up testing.

The third factor is ownership. It is not the intent of this chapter to discuss the relative merits of in-sourcing vs. purchasing chemistry. As the principles above apply to commercial compound production houses as well as pharma/biotech, the cost factors are similar with the exception of ownership. When compounds are obtained under varying right-to-screen agreements, the initial cost is usually substantially lower in lieu of downstream royalties and other delayed financial obligations. This allows relatively inexpensive entry into a larger collection, but the rights (ownership) of the compounds are not absolute, as in the case of internally synthesized compounds or compounds purchased under exclusive contracts.

Varying approaches can be used to assign a value to a collection. One source suggests that traditionally synthesized compounds be valued at $5000 to $7000 each, and combinatorially synthesized compounds be valued at $5 to $10 each (2). We prefer somewhat more conservative estimates. For internally synthesized compounds, where the average cost of a chemist and the required overhead is $250,000 per year, and where the average productivity is ~100 pure novel molecules per year in 15-mg quantities, the cost per molecule is then $2500. Using that figure to determine the value of the average big pharmaceutical library of 800,000 molecules suggests a value/cost of $2 billion. Most companies have purchased a large proportion of their library from commercial vendors. Depending on chemistry and exclusivity, these compounds generally cost ~$100 for 10 mg. That figure as a basis suggests $80 million for an 800,000-compound library. If the entire library could be produced using combinatorial techniques with a cost of $10 per compound, the 800,000-compound library would have a value of $8 million.

Since most collections employ a variety of sources, the cost of the collection will be some combination of the above factors. A cost of $500 million is a reasonable estimate. Clearly, the compound collection is one of the largest, if not the single largest, physical asset of any drug discovery organization.

4.1.5 Stewardship of the Corporate Compound Collection

The significant value of the compound collection, in terms of both the cost of acquiring the collection and the value it offers to the discovery organization in screening and lead optimization efforts, dictates that stewardship of the compound collection be carefully considered and managed. Usually the primary responsibility for stewardship of the compound collection is entrusted to the compound management organization. The collection must be used to best advantage and must be protected.

Protection entails physical protection from contamination, destruction by fire, or other catastrophe, as well as replenishment as parts of the collection get used up. This requires that the compound management organization set up adequate safeguards and processes to ensure that the collection survives into the future.

Using to best advantage includes setting up business processes so requests are fulfilled according to agreed upon service levels, and so no more than the minimum required amount needed be used. This requires constantly improving processing and usage procedures to deliver minimal amounts of compound more accurately.

Another part of stewardship involves assisting research in knowing compounds and services are available and can be used. There is little value to a collection that does not get used. In addition, there may be some "jawboning" needed when a researcher requests more than is reasonably needed for a specific step. Often, requiring special approval to order compounds in quantities larger than predefined limits provides a means of monitoring this.

Stewardship also requires establishing and maintaining the trust of the scientific organizations in discovery dependent on the compound collection:

- Compounds entrusted must remain available for their expected lifetime.
- Accurate inventory information must be maintained so requested compounds can be delivered and it can be understood why depleted compounds are unavailable.
- An effective reservation system must be set up so compound usage can be limited to certain areas while the compound is of intense research interest and then released to general availability at a later time.
- Unexplained disappearances of compounds should be minimized.
- Reasonable service-level agreements should be established, and the compound management organization should manage its affairs to ensure that those agreements are met.

4.1.6 Compound Sources and Collection Size

How big the compound collection needs to be is open for continued debate. Equally important are the sources, quality, information, and amounts for compounds. Is this emphasis on numbers just a game, or is the size of the collection of paramount strategic importance?

The growing emphasis on high-throughput screening has led to a requirement for increasing the size of the compound collection to improve the probability of getting quality hits from the screening process. As screening throughputs of 100,000 or more compounds per day through an individual assay become more common, it becomes reasonable, even essential, to screen large numbers of compounds. An initial screen may easily involve up to a million compounds, leading to a desire in many companies for large collections.

Conflicting with the desire for larger collections is the desire to perform screening more quickly and at lesser cost. Decreasing the number of compounds going through individual screening assays is an obvious way to decrease costs. With slower screening assays, particularly those more difficult to automate and carried out by the therapeutic area biologists in a more labor intensive manner, or those using very expensive reagents, there is a strong desire to screen only a portion of the collection. This problem can also be attacked by miniaturization, leading to lower amounts of compounds required and more effective automation, leading to higher throughputs.

Sources of compounds are also an important consideration. Many companies have built a franchise in specific therapeutic areas and chemical classes. Their internal collection is very rich in these areas. As companies try to extend their reach into new therapeutic areas and chemical classes, they need to extend their collection. It is not practical to assemble a chemistry staff to utilize traditional synthesis to quickly grow a chemical collection diverse enough to support all therapeutic areas and chemical classes. Various approaches are used to increase the size of the collection. Most companies use a combination of approaches.

Typical sources of compounds that are included can be found in Table 4.1.

Table 4.1 Sources of Compounds Found in a Typical Compound Library

Source	Major Characteristics of Source
Traditional internal synthesis	Provides compounds of unique structure and interesting properties. Company has complete intellectual property rights. Compounds are typically pure and synthesized in sufficient quantity. Most expensive.
Internal parallel synthesis, combinatorial chemistry	Provides large numbers of compounds of related structures. Compound purity and degree of characterization are usually compromised. Compound amount is often insufficient for follow-up, leading to need to resynthesize. Company has complete intellectual property rights. Least expensive until the costs of setting up and staffing the capability are considered.
External traditional synthesis, purchased compounds; includes compounds from universities and foreign institutes	Compounds of unknown quality need effective quality control (QC) protocol. Compounds typically pure and available in decent quantity. No or weak intellectual property rights. Other companies often have the same compounds.
Contract synthesis, both traditional and parallel synthesis	Typically same characteristics as comparable internal source. Usually includes full intellectual property rights. More expensive.
Purchased combinatorial libraries	Unknown quality. Often based on interesting chemical structure the external company specializes in. Variable intellectual property rights. More expensive than internal parallel synthesis.
Shared combinatorial libraries—obtained from company specializing in this business	Unknown quality. Often based on interesting chemical structure classes of interest to several companies. Often do not know chemical formula and structure. Limited intellectual property rights. Least expensive. Must consider whether screening and follow-up can occur quickly enough to gain intellectual property rights and structure information if desired.
Natural products, internal and external	Complex mixtures of naturally occurring substances. Must be processed/separated into smaller mixtures. Screening hit rate is often high and dominated by materials difficult to isolate and characterize. Needs special staff skills to isolate and characterize. May not be able to synthesize directly, even if structure known. Difficult intellectual property issues. Expensive but rich source of hits.
Collection exchanges	Viable between noncompeting companies, such as between pharmaceutical and agricultural products companies. Compounds may not be of much interest because of different characteristics of the industry's products. Difficult intellectual property issues to resolve.
Partnerships with smaller companies	Collections typically limited to a particular activity type or synthesis method. May include exclusive intellectual property rights. Good way to expand the diversity of a collection in a strategic direction.

Other important considerations include the ability to utilize subsets of the collection for various purposes and the nature of the lead optimization capabilities. A common approach is developing the ability to screen a diverse subset of the entire collection followed by assembling another subset containing compounds structurally related to the hits from the diverse subset. Depending on the depth of the collection and the hit-to-lead strategy of the company, this may be followed by internal or contract synthesis of a large number of structurally related compounds using parallel synthesis techniques, which then use quantitative structural-activity relationship techniques before starting the search for the best compound in detail.

4.1.7 Dependencies on and from Other Industrialization Activities

As stated earlier, compound management has grown in response to changes in related and dependent activities, such as high-throughput biological screening. The explosion of genomic data coupled with numerous technology advancements has led to the wide adoption of a statistical approach to drug discovery. This presents a particular challenge as screening strategies and technologies continue to evolve.

Genuinely industrialized processes have the common elements of standardization and reproducibility. The detailed specifications of a 4-inch clay flowerpot do not vary much over time. This allows the establishment of a standard manufacturing approach, which, while very effective, is seldom flexible. In drug discovery, screening paradigms are constantly changing, but the required output numbers remain very high. This leads to the need for constant vigilance and monitoring for change. An apparently small change in required sample format can completely derail an otherwise successful process. Said differently, there is always a trade-off between efficiency (throughput and quality) and innovation (flexibility). The management of this dichotomy is key in keeping an initially successful process successful for the long term.

The true keys to the successful management of this issue are coordination and communication. Sample formats will change, but the rate of change can be carefully controlled via communication and coordination with the customers of the compound preparation process. If the changing needs of the screening groups are shared with the compound room early enough, a successfully coordinated change management effort can be achieved. Vigilance in managing these relationships is essential.

Another key dependency is the approach to screening program life cycle. Understanding how many new screens will be added per year vs. the number of older screens that are retired is essential. Adding new screens without retiring (or revalidating older screens) produces the unfortunate, but common side effect of additive change. For example, it is frequently the case that when a new sample format is added, no other formats are retired. The apparent benefit of this type of change is that a wider variety of sample formats are available for screening, but the increase in costs, in both dollars and compound consumed, can be substantial. This can be seen when the preservation of older equipment and processes is required solely to maintain the capability of delivering a single-sample format. The specific costs include the maintenance costs of the equipment; single-purpose units and additional compound that is consumed making these samples consume lab footprint.

4.1.8 Value Proposition

It is not expected that a new molecular entity (NME) will be found in the compound collection. A diverse collection will allow a screening organization to discover one or more leads. Several cycles of lead optimization by the chemists synthesizing new compounds will hopefully result in an NME. Since compound logistics processes do not produce NMEs on their own, the value of this function or collection will always be the subject of debate. Many organizations desire to direct their investments toward increasing the size or diversity of their collection, rather than investing in the infrastructure necessary to optimize the utilization of the collection. There is no right size for collection size or infrastructure budget.

Traditionally, it has been difficult to justify large investments in compound management infrastructure. That said, virtually all organizations with large screening programs have made these investments. Typical justification factors include "projectizing" the compound management infrastructure with biological testing investments and decreased discovery cycle times by removing, or never creating, compound management bottlenecks. In a sense, time is money; therefore, time saved is money earned. It is even more difficult to establish budgets for maintenance activities such as resolubilizing compounds.

It is reasonable to set some general guidelines. For a large investment in a compound collection in the magnitude of $1 billion, it is reasonable to expect a significant expenditure on the order of $20 million for the infrastructure to make utilization of the collection possible. Further, since solubilized compounds are gradually used up or degraded in storage, an operating budget to keep the collection in usable collection and to develop new methodology is needed, perhaps $2 million per year.

This is analogous to the thinking when building a new manufacturing plant or research center. The large capital expenditure generally includes the expectation for a significant capital expenditure on IT infrastructure and software, and then ongoing annual maintenance of the IT infrastructure is usually budgeted at 15 to 20% of the original cost.

4.1.9 Basic Capabilities of a Compound Management Organization

There is no one-size-fits-all description of the capabilities of the ideal compound management organization. It must fit the needs of its discovery organization, but beyond this, it is very dependent on the goals and organization of the specific company. A few generalizations are useful in defining the basic requirements for compound management in all research organizations (see Table 4.2).

Table 4.2 Basic Requirements of Research Compound Management Organizations

Capability	Description and Discussion
Manage collection	The size and characteristics will vary to suit the needs of the organization. Ability to store, protect, maintain, and improve the collection. Includes improving diversity, acquiring new compounds, and maintaining integrity of inventory data. Ability to replenish compounds and solutions as they are exhausted.
Make collection accessible	Ability of research scientists to know the makeup, structure, amount, and location of compounds in the collection. Ability to request individual compounds, the entire collection, or specific collections in useful formats.
Ensure compound quality	Provide quality information about compounds in the collection and techniques used in storing and delivering compounds so organization can move forward with confidence and efficiency.
Promote efficient compound utilization	Utilize stewardship techniques and advances in technology to ensure only minimal amounts of compound are consumed to meet the needs of the organization.
Prompt and efficient service	Establish and meet service-level agreements for all services. Provide services less expensive and faster than customers could service themselves.
Flexibility and automation	Ability to anticipate demands for new services and provide leadership in new services and automation. Ability to respond to changing and differing needs of HTS organization, chemists, and therapeutic areas.

4.2 Technology and Processes in Compound Management

Building or reengineering the technology and processes for an effective compound management organization is a complex and daunting task. It requires vision, resources, and talent as well as a certain amount of serendipity and luck. A major consideration is designing technology and processes to be both modular and high volume. This enables efficiency, automation, and training for specific tasks while also allowing flexibility for future enhancements. Technology and processes are typically separated to support the following activities:

- Neat compounds.
- Compound solubilization.

- Storage and distribution of solubilized compounds on custom plates and in individual containers in various formats and concentrations.
- Library replication.

4.2.1 Neat Compounds

The most fundamental and long-standing process in compound management is the neat compound process. The term *neat* refers to single, relatively pure samples that exist in dry form, i.e., without solvent or other diluent. Typical neat processes involve the submission of neat samples to the compound room by chemists, the short- and long-term storage of these samples, and the eventual distribution of portions of these samples to biologists for testing.

4.2.1.1 Storage Conditions

Since the compound collection is often considered the "crown jewels" of a research organization, it is imperative that the integrity of the collection be maintained. Each sample or sample mixture container must be uniquely identified. The location and amount of sample in each container must be known at all times, and the conditions in which each container is stored must be optimized to reduce sample contamination or degradation.

The ideal neat sample storage conditions have historically been a subject of debate. There is general agreement on what the key variables are—heat, light, temperature, oxygen exposure, and moisture/humidity—but there are differing opinions on which are most relevant and what the best practices should be. These issues are best discussed locally in each organization, but a few general rules should be applied.

As many stable compounds can be affected by light, samples should be stored in the dark. This can be accomplished by using amber-colored sample vials, having a dark storage area, or both. Humidity and temperature effects can usually be considered together, and humidity is typically the greater concern. Many have argued that drug-like molecules, especially those intended for eventual oral administration, should be stable at ambient temperatures and therefore refrigeration (4°C) or freezer (–20°C) storage of neat samples is unnecessary. This position can be considered valid as long as the relative humidity is controlled. In the northeastern U.S., for example, there are significant seasonal shifts in ambient temperature and relative humidity (RH). Building air conditioning typically controls temperature variations, but even in tightly closed vials, samples can take up water during times of high RH and can give up water during drier winter months if humidity is not tightly controlled. This water uptake can potentially change the character of some molecules, and it can also lead to disparities in the weights of samples and containers. For these reasons, humidity control is essential. Protection from oxygen and storage at reduced temperatures are less often employed.

Many companies keep some of their compounds at reduced temperatures, in dessicators, or in the absence of oxygen. Although drug-like compounds should be more stable, it must be remembered that compounds in the collection are not intended to be the next big drug: they are hoped to display activity in a high-throughput screen and encourage chemists to synthesize the next blockbuster based on activity and structural properties.

Protection from fire is a significant consideration. Because of the high value of the collection and also the value of the facilities used to store this collection, insurance is often used to defray the cost of replacing the collection in the event of a disaster. Along with insurance, there is the need to extinguish a fire if one occurs. Compounds in the collection are organic; thus, most will burn under appropriate conditions. The storage of flammable materials is generally accompanied by a requirement for a fire protection system. This system may be as simple as a water-based sprinkler system or a more sophisticated system to provide a blanket of nonflammable gas to extinguish a fire. Systems of the latter type are very expensive and of limited utility for many facilities. In the event of a fire and a "dump" of the nonflammable gas, the storage area must be quite tightly enclosed to keep the inert gas contained as long as the heat source continues to exist. Since the most likely source of a fire is electrical failure of components of the storage equipment, this blanket of inert gas must persist long enough for something like an overheated motor to cool off.

4.2.1.2 Storage Containers

As stated previously, it is important to protect samples from light, heat, and changes in temperature and humidity. A well-selected neat sample container is the first line of defense. The protection from light has often led to the decision that neat sample vials should be amber. I have found that this is not necessarily the case. If the storage areas are dark, the samples will not be exposed to light for extended periods. This leads to the consideration that the vials themselves can be clear. It is easier to see and weigh small samples in clear vials, especially very small samples, which must be scraped from the vial walls. In addition, many of the chemicals were not synthesized in the absence of light and therefore should not be extremely photosensitive. As with all things, this may vary depending on the composition of any given compound collection.

Ideal sample containers should also be designed to minimize variations in the vial tare weight. The issue here is often a weight variation caused by uptake and release of water from paper label/bar code stock. In may be counterintuitive, but slight changes in vial label weight can raise havoc with sample weight integrity systems.

Vials with internal round or v-bottoms are a nice feature when working with sticky compounds and using solubilization and split techniques. These have the advantage that pipet tips can capture the solution with minimal dead volume.

4.2.1.3 Weighing Technologies

Sample weighing is the basis of most compound management/logistics processes. Samples are typically check-weighed into a compound room. They are usually weighed again to be aliquoted into small amounts for delivery to biological testing, and accurate weighing is the cornerstone of high-quality solubilization activities. Lastly, samples are typically check-weighed prior to return to storage after processing.

In truth, there have been very few fundamental changes over the years in balance and weighing technology used in compound management. For high-quality, low-throughput applications, a human sitting at an analytical balance using a spatula or similar device to weigh individual samples is still the basis of most compound rooms. The balances themselves can vary a bit in precision and bells and whistles, but the basic user experience has not changed much. The basics of this manual approach require balances that are accurate and precise, are calibrated, maintain their calibration once obtained, quickly stabilize to a weight reading, have a housing that reduces static electricity issues, and are easy to read, operate, and clean.

There are two primary reasons why the manual approach is still preferred. First, the physical state of materials submitted into compound management processes is highly diverse. In addition to the ideal of clean and pure free-flowing white powders, there are often samples that can be more appropriately described as tars, oils, and the like. These types of samples do not readily come free from their containers, stick to handling implements, and generally foul up automation equipment. The second reason is simply economics. To date, automated strategies are simply not cost-effective and reliable compared to the more flexible human.

The difficulty in dealing with sticky compounds and extremely small samples is leading to solubilize-and-split techniques as an alternative to traditional scraping and weighing. If the amount of material in a container is known, then the entire sample can be dissolved in a known amount of a volatile solvent, followed by transferring an aliquot to a dispense container. Solvent can be removed under vacuum from both containers. The storage container can be check-weighed and the amount in the dispense container can be calculated by difference, from the volume proportion of the aliquot, or by check-weighing the dispense container. This solvent transfer process involves more equipment and steps, but virtually eliminates the waste and frustration in dealing with sticky compounds. A complication is that most computer systems that support the manual weighing process need to be modified to support a solvent transfer process.

4.2.1.4 Neat Compound Automation Strategy

Storage systems for neat sample vials range from completely manual systems employing shelves and drawers, to semiautomated systems that involve automated shelving systems but have a person pick and place vials, to fully automated systems that receive a list of desired containers and produce a rack or tray of desired vials without human intervention.

Manual storage systems have the advantage of lowest cost and maximum flexibility to accept a variety of container types and sizes. Major disadvantages are that the entirely manual system is prone to operator error, resulting in lost or misplaced vials, and that manual systems are relatively limited in size, often for safety reasons, resulting in a relatively low density of storage per square foot of floor space. From a cost standpoint, there is a trade-off between cost of the equipment and cost of the ongoing operations. With a manual system, the cost of people to manually pick and place and hunt for missing samples must be considered.

Semiautomated storage systems typically involve movable shelves in either a revolving carousel arrangement or slide-out robotic mechanism. In either case, shelves can be arranged with minimal lost space between shelves and can be stacked very high, taking maximum advantage of floor space. Flexibility can still be maintained by having different shelves at different separation distances. The cost is intermediate between costs of manual and fully automated systems. An attractive feature is that with some forethought, additional robotics can be crafted to facilitate automated picking and placing at a later time if needs change or additional expenditures can be justified. Ongoing operational labor costs are somewhat less than those with manual systems, but still a major factor.

Fully automated storage systems have the major advantage of virtually eliminating missing vials and greatly reducing ongoing operational labor costs. The cost of a fully automated system is significantly greater. Storage density is comparable to that of semiautomated systems. Flexibility is generally limited by the ability of the automation to accept only one container type.

Fully satisfactory automated powdered compound dispensing and weighing systems for handling neat samples have not been very effective to date. Automated weighing systems that mimic the actions of a human operator, including removing the vial cap, taring the receiving vial, shaking a portion of the compound into the container using a shaking/vibration technique, and weighing both the receiving and dispense vials, work for a reasonable percentage of samples that are free-flowing powders. However, most collections contain a significant percentage of difficult-to-handle compounds. In addition, the automated systems usually produce a number of vials that miss the targeted dispense weight and the vials must be set aside for manual adjustment. This lengthens the overall process, eliminating most of the automation benefits.

Another approach to powder dispensing utilizes a special vial cap containing an auger mechanism that doses out specific amounts of dry material. This has the potential to work well with some compounds but introduces additional problems. If a separate special vial cap is affixed to every vial, this approach will be very expensive for large collections. If the special vial caps are placed on the vials at the time of dispense, then there are potential problems with contamination, cleaning, and drying of the caps.

The potential savings from an effective automated powder compound dispensing and weighing system are quite large because the weighing and dispensing steps are the most time consuming activities in a typical compound management organization. However, because of the problems cited above, the authors have little optimism that such a system will be developed.

Neither of the above approaches can be used for sticky compounds, which comprise the majority of compounds prepared using parallel synthesis.

Another approach to neat compound dispensing creates solutions using specific volatile solvents followed by aliquoting of the solution to dispense a specific amount of material. After removing the solvent under vacuum, the amount of material can be calculated by using the ratio of dispensed volume to dissolved volume, or alternatively, the containers can be weighed to determine the amount left and the amount transferred. This approach is attractive because it is simple, can be automated, and employs techniques used by chemists in the preparation and purification of compounds. The negatives to this approach include the cost of the equipment to remove solvents effectively, the difficulty in working with multiple solvents, and the uncertainty regarding the long-term effects of multiple dissolution and drying operations.

4.2.1.5 Example Neat Compound Weighing Business Process

An example of a traditional weighing process currently in use in industry is provided in Figure 4.1. This figure shows an example process that combines physical sample movements/manipulations and electronic data flows. The flow assumes that the sample of interest has been submitted previously into the system electronically and physically by any number of sources. The flow follows the process from sample request through fulfillment processing and delivery to the requestor.

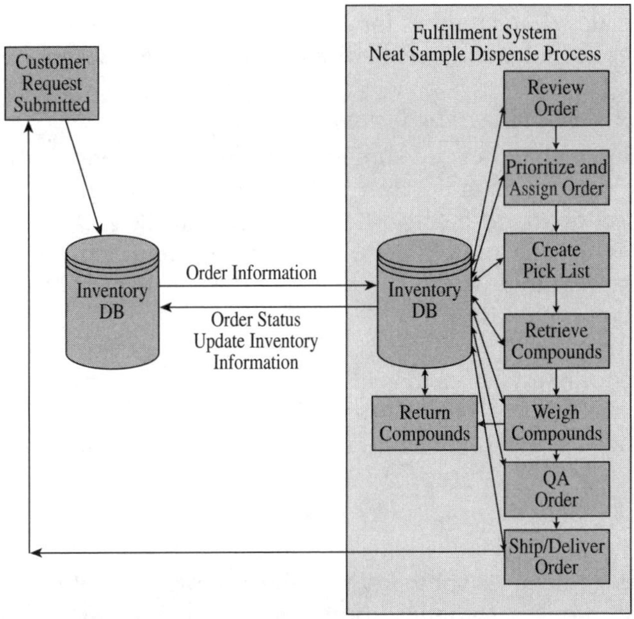

Figure 4.1 Example of the neat sample dispense weighing process.

4.2.2 Compound Solubilization and Storage

Compound solubilization is the process of dissolving compounds into solvents so they can be delivered as solutions for screening and activity testing. Depending on the required concentration tolerances for the desired solution, the capacity of these processes can range from low to high throughput. There are two general types of solubilization processes: custom solutions and microtiter plate-based solubilization.

In the first case, individual compounds are weighed into containers and various solvents added to produce individual solutions of specified final concentration and volume. This is typically a low-throughput manual operation that provides custom solutions to support specialized *in vivo* or *in vitro* testing, but not high-throughput screening.

In the microtiter plate-based process, the goal is to produce large numbers of individual solutions of uniform concentration and volume within a plate. In this case, solvents are usually identical across a plate or rack of tubes as well as across multiple plates. The most common solvent is DMSO. Variations on the plate production approach can include mixtures of compounds or pooled samples within the plate wells.

The name of the second approach is often a misnomer because in many companies the target container is not a microtiter plate but a set of tubes or plates.

The general approach used to support screening at most companies is to solubilize large numbers of compounds as they are received in the compound management organization. The solubilized compounds are stored in deep-well microtiter plates or microtubes for use at a later time. These master plates or source plates or microtubes are then stored and used in subsequent processes.

4.2.2.1 Solubilization Techniques

Solubilization techniques are similar for both manual and automated processes. Dry compound and solvent are mixed together in a container until solubilization occurs. If a solvent mixture, such as 30% DMSO in water, is desired, it is usually preferable to dissolve the compound first in DMSO, then add the required water. This is usually faster than adding the mixture to the dry compound directly.

Many compounds are sparingly soluble. It is often necessary to shake the compounds for many hours to get them to dissolve. Application of some heat will speed the process, but this may also accelerate decomposition, so it is done sparingly. Use of ultrasonification to promote solubilization is common.

4.2.2.2 Solubility Variables and Solvent Selection

Compound solubility is a tricky component/variable in large-scale compound management. As solubility is a physical property of individual compounds, variations in solubility should be expected across large compound collections. Wide variations in solubility are experienced in different solvents. What is needed is the ability to utilize the entire collection in useful concentrations in a solvent suitable for an assay. Almost universally, the method used to manage the solubility variations seen across large libraries is the use of a near-universal solvent such as DMSO. No other solvents have proven to have wide applicability in a variety of screening assays. In combination with setting a top concentration of 3 to 10 m*M*, this makes handling a large number of compounds in solution practical.

For use in screening large libraries, solvent selection is usually a function of the intended chemistry and biology of the drug discovery and is optimized to ensure accuracy and repeatability in the screening process, as well as long-term stability of the wet compound collection. DMSO or a DMSO–water solution is generally the only solvent considered for libraries intended for HTS screening. (A useful reference on the chemical, physical, and handling properties of DMSO is the *DMSO Handbook* (3).)

Some assays, particularly cellular assays and *in vivo* testing cannot tolerate DMSO. As solubility is a physical property of individual compounds; the issue of solvent selection can be involved. Detailed solvent selection for individual compounds and assays typically occurs downstream of HTS in the lead optimization and secondary screening process, where custom formulations are designed to improve the bioavailability of drug candidates. The details of this process are best left to a formulations text.

4.2.2.3 Storage Conditions

Solutions may be stored at room temperature, refrigerated typically to 4°C, or frozen at –20 or –80°C. There are advantages to each approach. In general, it is believed that lower temperatures lead to slower chemical decomposition reactions. However, there is some evidence and speculation that decomposition is accelerated at the liquid–solid interface in solutions as they freeze and thaw. There is also considerable experience that freeze–thaw cycles are bad, regardless of reason, and some organizations set an arbitrary limit to the number of freeze–thaw cycles before solutions are discarded. No universal rules have been accepted across the industry. Another reason to avoid cooling and freezing is that the solubility of most compounds decreases with the temperature, causing the compounds to precipitate. Once precipitated, some chemicals are difficult to get back into solution. It is not uncommon to observe precipitated material in storage containers.

Storage at room temperature has the advantage that it is easy, cheap, and allows sparingly soluble compounds to stay in solution. The evidence that lower temperatures are beneficial is debatable.

Storage at 4°C generally gets the solutions to solidify because this is the temperature at which DMSO freezes. If 4°C is good, most people feel –20 or –80°C is better. Unfortunately, such generalizations are hard to prove. Different people will rank various factors more important. Also, different combinations and purities of compounds can lead to different conclusions.

Regardless of the storage temperature decided upon, there are other storage conditions to be considered. These include humidity, light, and exposure to oxygen.

Storage containers such as microtubes or microtiter plates utilize seals to protect their contents from moisture and possibly oxygen. However, these seals sometimes prove unreliable, and it is often desirable to provide a second level of protection. For this reason, humidity is usually carefully controlled where compound solutions are stored. This is because DMSO acts as a strong desiccant.

Fire protection is another consideration for solution storage systems. The containers—microtubes or microtiter racks—are usually made of plastic and are flammable. DMSO is a flammable organic solvent, although the flash point is quite high. Even the compounds are flammable. Thus, some sort of special fire protection is usually required. If the storage is in a cold box, particularly with internal robotics, a sprinkler system should be used as a last resort. Even if the fire department or insurance company requires a water sprinkler system, it is desirable to install a primary fire protection system that utilizes a nonflammable blanket of gas and only activate the water system if the primary system fails to extinguish the fire.

4.2.2.4 Storage Containers

Compounds dissolved in DMSO are most commonly stored in microtubes or microtiter plates. Compounds dissolved in other solvents may be stored in vials with Teflon-lined caps or in microtubes or microtiter plates.

The volume of a microtube is often about the same as a deep-well microtiter plate. The decision as to whether to store in tubes or plates is generally made based on how the compounds are to be used. It is quite common for organizations to have some of their inventory in tubes and some in plates.

Microplates typically have a different compound in each well. Microplates are most efficient when all the compounds on a plate need to be used or treated the same as when they are part of a library for primary screening.

Microtubes allow a single container to contain only one compound. A system used to produce custom plates would ideally use microtubes because then only the compounds needed would be retrieved, rather than having to handle all 96 compounds on a plate even if only 1 was needed.

Microtubes may be stored in 96-position microtube racks that are stored on shelves. Microtubes may also be stored in large trays that are stored on racks. The latter generally allow a higher storage density and may be more precise for robotic handling. Microtube racks are good if it is sometimes desirable to handle the tubes in sets—then the robotics can be considerably faster.

Sealing of both microtubes and microplates is an important consideration. The primary purpose of sealing is to avoid contamination with water or adjacent compound solutions and to avoid evaporation. Of these, preventing uptake of water provides the most stringent requirements.

Microtubes may be sealed with plastic caps, piercable septa, or piercable silicone plugs. Usually, any of these techniques provide reliable seals.

Use of plastic caps does require that the caps be removed to withdraw aliquots of the solution. This causes the solution to be exposed to moist air during the period the cap is removed and also leads to challenges in automation in the form of a capping–decapping mechanism.

Piercable silicone plugs tend to produce the most satisfactory seal, but few robots are capable of piercing the plugs reliably. Piercable septa are easier to use with automation, but long-term experience with the quality of the seal is not available yet.

Microplates can be sealed with cap mats, adhesive seals, or heat seals. Cap mats are easily applied manually with a roller or press. Automated placement or removal is difficult. Adhesive seals are easily applied manually or with automated systems. Seals can usually be easily removed and fresh seals reapplied. Although seal removal is easy manually, automation of this step is difficult. The adhesive coating is sometimes a source of contamination and must be carefully evaluated. Different adhesives perform differently under differing temperature conditions. Only a few adhesives are effective at –80°C. Removal of an adhesive seal leaves behind a small amount of adhesive on the top of the microplate. This adhesive may cause plates to stick together and can cause problems with automation. Most adhesive seals are not reliable for long periods. DMSO vapors can attack most adhesives, causing leakage and a sticky mess.

Heat seals use a special sealing material that is fused to the top of the plate. Since the fusing involves some melting of the top of each well, some deformation of the plate occurs upon heat sealing. Although this deformation is usually minor, it does limit the number of times a plate can be resealed. Since the seal is actually fused to the top of the plate, seals can be difficult to remove. Some sealing materials are designed to be peelable, but removing the seal can be difficult and is difficult to automate. Heat seals have been found to be good for storage at low temperatures and generally are believed to produce the best long-term seals. Some companies use piercing devices that place a small hole above each well to allow automatic pipettors to access the wells. If tips touch the seal material, there are often problems with the automation inadvertently moving plates around.

Regardless of the type of containers or type of seals used, the materials must be carefully evaluated for their ability to contaminate DMSO solutions. Many plastics contain antioxidants or plasticizers that can leach into the DMSO and interfere with assays.

4.2.2.5 Solubilization and Storage Automation Strategies

The automation strategy is best divided into two parts: automation of the solubilization process and automation of the storage process. There is little need to provide an integrated solubilization and storage process because they are two distinct steps.

4.2.2.5.1 Solubilization Automation. Compound solubilization is often rate controlled by the weighing process, which is often the first step. As previously stated in the neat compound section, compound weighing remains primarily a manual task. After compounds have been weighed into solubilization vessels, usually pre-tared test tubes, a specific volume of solvent is added. Once the solution is adequately mixed, a solution of specific final volume and concentration is obtained.

The solubilization step is conveniently automated using a liquid handler. These machines vary widely in cost, capability, and complexity. The liquid handler must be capable of adding solvent to the neat compound and, after solubilization is complete, transferring the resultant solution to the appropriate storage containers. In most cases, these units simply mimic their manual cousins, the hand pipet. Automation eliminates the tedium of the manual process, eliminates most mix-ups or mistakes, and significantly increases the throughput and reduces costs of the solubilization process.

The most common challenges in any solubilizaton process are the weighing steps and the stability of the weighed materials, the inherent solubility (or lack thereof) of the target compound(s), and the accuracy and precision of the liquid handling device. The challenges in weighing steps have previously been discussed. Exposure to the environment, specifically to light and excessively low or high humidity conditions, can bring additional problems. Some solubilization processes can be lengthy, and this can allow the compounds to absorb or desorb water, which effectively can change the mass of the sample. For these reasons, the optimal conditions of low light and controlled temperature and relative humidity are essential for the lab space as well as the storage areas.

Liquid handlers come in a variety of shapes, sizes, costs, and capabilities. Most laboratories prefer to pick a specific liquid handler for a given task and dedicate it to that task. General-purpose liquid handlers are also available that can perform multiple tasks, but these are often more expensive and usually do not perform all of the desired tasks equally well.

The following capabilities need to be considered for a liquid handler used in the solubilization process:

- Ability to work with relatively large volumes of solvent, i.e., 1 ml.
- Ability to dispense variable volumes into each container based on input from a file or another computer system.
- Ability to minimize carryover through changing tips or effective tip washing.
- Compatibility with different types of containers and layouts within the same procedure—ability to dispense into racks of vials or test tubes and later transfer to plates or racks of minitubes for storage.
- If the organization stores solubilized compounds in minitubes with septum-type closures, special consideration needs to be given to the ability of the liquid handler to pierce the septum.
- Compatibility with tube and plate sealing processes and procedures for dispense containers.

4.2.2.5.2 Storage Automation. The storage automation strategy of solubilized compounds must consider the container types to be stored, the storage conditions, how the containers will be used, and the degree of integration needed with equipment used for dispensing solubilized compounds. Storage automation is usually designed to work with the automation equipment that will dispense liquid from the storage containers.

Storage systems for solubilized compounds are analogous to those used for neat compounds. Manual, semiautomated, and automated systems can be used, and it is common to utilize a mixture of these approaches.

Several vendors make automated tube stores. These systems range in capacity from a few hundred thousand microtubes to several million microtubes. Tubes are stored in large trays, sometimes in microtube racks of 96 tubes. In some automation fashion, a list of desired tubes is submitted to the system, and the system moves the trays containing the desired tubes to a picking area. In the picking area, the tube will be transferred by a robot into a rack of desired tubes. When all the tubes on the list have been picked, the racks containing the desired tubes are moved out of the storage system to a place where they can be transferred to a dispensing workstation. In fully integrated systems, these racks may be physically moved using a conveyor device and not requiring any user interaction. To be considered an automated tube store, the unit must at least have a mechanism to remove and store microtubes without an operator. Some vendors do not require a robot to do the pick and place, but have a pneumatic or other unique device. (Specific vendors are not mentioned here because the list of vendors is large and constantly changing. Several useful scientific meeting sites maintain vendor listings and scientific paper information (4).)

Semiautomated tube stores are analogous to the vial stores. The shelf containing the tubes of interest is moved to a position where an operator can pick and place the tubes.

In a manual tube store, the tubes would typically be stored in racks located on shelves. The operator would manually locate the rack and then the tube.

Storage density is an important consideration for all tube stores. The maximum density is usually obtained with trays that contain a few thousand microtubes each. These systems are difficult and tedious to use without fully automated pick-and-place robots.

The decision as to how much automation to utilize is usually based on how many tubes are accessed each day. If a tube is only accessed once or twice, and fewer than a hundred tubes are accessed each day, a manual system is satisfactory. If several thousand tubes must be accessed each day, a fully automated system is justified.

Similarly, microplate or microtube rack storage systems can be manual, semiautomated, or fully automated. Fully automated systems can also be integrated to pass the plate or rack from the storage system to the solution dispensing system without operator interaction.

The daily volume of plate access is rarely sufficient to justify a fully automated plate storage system. A fully automated plate storage system is usually only used in those systems where full integration with a dispensing system is required and controlling the amount of time a plate is unsealed and at room temperature is of paramount importance.

Some additional concepts of storage automation are discussed with the automation of the compound solution dispensing systems.

4.2.3 Custom Solutions

The preparation of custom solutions was mentioned in the introduction to this section. This is generally a low-throughput process and the steps are generally manual. Concentration ranges and solvents used are not standardized as with solutions used in support of screening. In many companies, custom solutions are not provided by the compound management organization, but are made in the user organizations. In this case, the scientist will make a neat request to get the compound and make the solutions as needed.

Custom solutions are usually only provided to support the nonscreening processes of discovery. Solutions are typically used in either *in vivo* or *in vitro* tests. Often, these tests cannot tolerate DMSO and a variety of solvents may be employed. These solutions are provided as a service so biologists do not have to spend their time making solutions. In this process, the compound management staff must be fully trained in chemical solution concepts and must be prepared to deal with unusual requests and situations.

The requesting and preparation of custom solutions is often not precise because the solubility of research compounds in various solvents is not documented and often not known. Often, the staff must first determine the solubility and occasionally search for a suitable solvent for the test the researcher is interested in performing. The compound management staff will usually maintain a database of measured solubilities in popular solvents. Occasionally, this information is accessible by the scientists requesting solutions, but often it is not. In some companies, the synthetic chemist is expected to provide some solubility information as part of the compound characterization.

When custom solutions are made, there may be more solution made than needed to fulfill the request. This stock solution can be saved as an inventory item and used for subsequent requests.

4.2.4 Solubilized Compounds

Solubilized compounds are prepared and used within the compound management organization to use for libraries and the source material for custom plates. The solubilized compounds are typically stored in deep-well plates or microtubes, although other containers are possible.

4.2.4.1 Libraries

All the compounds a company has available are usually referred to as the collection. Not all compounds are suitable for screening. Compounds suitable for screening are generally grouped into one or more libraries. A library is loosely defined as a group of compounds. Generally, it is common for a compound to belong to more than one library.

Classification and nomenclature of libraries differ at each company. Each company needs to define libraries in a manner to support its screening strategy and make screening easier and more effective. Most companies have an internal protocol to define libraries, to group compounds into libraries, and to use the libraries in screening. The characteristics important in defining a library might include date compound first available, structural features, source, activity types, and so on. It is common to include nearly all compounds in a master library, to define a subset of that library, including the most common structural and activity features as a diverse library, to define all the compounds received between certain dates as an update library, to define all the compounds possessing certain structural scaffolds into separate combinatorial libraries, to define all the compounds likely to exhibit specific activities (i.e., kinase or protease) as class libraries, and to define sets of compounds purchased together as distinct purchased libraries.

Libraries are generally delivered to screeners on one or more microtiter plates. All compounds on the plate should be present in the same solvent and concentration.

At many companies, screening can be an ongoing process, meaning that after initial screening of available libraries is complete, as additional compounds become available, they are presented to the screen. Because of the time and cost of screening, it is important that the protocols and procedures followed make it straightforward to identify, request, and deliver new compounds to the screen without delivering previously screened compounds.

4.2.4.1.1 Library Replication. When a library is needed for screening in one or more assays, the compounds must be provided to the screeners in a suitable format. Some screening organizations want assay-ready plates, that is, plates that can be used directly in the assay without further dilution or solvent changes. Other screen organizations prefer to have intermediate master plates that they can access several times to make their own assay-ready plates.

For a large pharmaceutical company there may be multiple screening organizations at multiple sites, each having responsibility for several HTS assays. Different assays may have different automation or requirements. It is common to require the flexibility to provide copies of a library in 96-well, 384-well, and 1536-well plate formats. It is also common to provide plates in more than one DMSO concentration and more than one compound concentration. In addition, some companies will prepare copies of the library in multiplexed format.

For maximum efficiency and lower cost, most compound management organizations prepare the library plates in mass production campaigns, attempting to make all the plates of a given library that will be needed during the next several months. This means they need the capability to make multiple copies during the campaign. Usually this requires specialized plate replication equipment.

The compounds are usually stored at a relatively high concentration (3 to 10 m*M*) and need to be diluted substantially (to at least 0.3 m*M* and often much lower) for screening purposes. Each company will have favorite storage and assay concentrations, and there is no industry standard in this area.

The compounds in a library can be stored in microtubes or microtiter plates. It is probably too confusing to use a mixture of tubes and plates for a given library. In general, it makes most sense to store large libraries in plates and small libraries in tubes. It is common to use 1.2-ml microtubes or 1- or 2-ml deep-well microtiter plates containing 96 wells to store library compounds for use in screening.

The possible complexity of the replication task is best illustrated through an example. The primary screening library for a major company typically contains 500,000 compounds. This would be about 6250 master plates containing 80 compounds each. In a possible replication scenario, it might be necessary to make 18 copies of the library in different plate densities, compound concentrations, and solutions. For example:

- Two copies in 100% DMSO in 96-well plates at 0.4 m*M* concentration
- Two copies in 100% DMSO in 96-well plates at 0.04 m*M* concentration
- Four copies in 30% DMSO in 96-well plates at 0.4 m*M* concentration
- Two copies in 30% DMSO in 96-well plates at 0.04 m*M* concentration
- Four copies in 30% DMSO in 384-well plates at 0.4 m*M* concentration
- One copy in 30% DMSO in a 384-well plate at 0.04 m*M* concentration
- One copy in 100% DMSO in a 384-well plate at 0.4 m*M* concentration
- One copy in 100% DMSO in a 384-well plate at 0.04 m*M* concentration
- One copy in 100% DMSO multiplexed five compounds per well in a 384-well plate at 0.4 m*M* total combined concentration and 0.08 m*M* individual concentration

4.2.4.1.2 Library Replication Automation Strategy. Liquid handling equipment used in replication is often dedicated to a single purpose because of the workload involved and importance of the task to the discovery organization. Characteristics of replication work cells include:

- Considerable storage for the source plates, destination plates, filled plates, and tips
- Integrated plate label module
- Integrated plate sealing module
- Ninety-six-well head
- Second head for solvent addition and mixing
- Capability to utilize 96-, 384-, and 1536-well plates
- Capability to produce multiplexed plates in the same production run

Creation of plates from compounds stored in tubes usually requires an additional step. In this case, the tube storage and custom plate system is typically used to transfer sufficient solution from each tube in the library set to meet the needs of the replication campaign. This solution is generally transferred into microtiter plates that can then be used in the replication work cell described above. If the throughput requirements are not too large, an alternative is to transfer the solution from the microtubes and make the destination plates using an eight-channel liquid handler capable of piercing the closures on the microtubes.

It would be attractive to store everything in microtubes and not have to deal with storage, retrieval, and sealing of master plates. However, throughput limitations of current tube storage systems make this impractical. Current systems are typically limited to retrieving and replacing 3,000 to 10,000 tubes per day. For a primary screening library of 500,000 compounds, this would require over 50 days' throughput capacity for each replication campaign and would take the custom plate capability out of use for that time. For this reason, microtube storage is currently only practical for small- to medium-size libraries.

4.2.4.1.3 Library Replenishment. Master plates typically have enough dissolved compound to last a year or two. Some companies make multiple copies of the master plates or master tubes, so their supply may last longer. Regardless, eventually the inventory of dissolved compound will be exhausted. At this point, replenishment must be considered.

If additional neat compound exists, an additional portion of each compound in a library can be solubilized. However, with each solubilization of the collection, some of the compounds will have their neat inventory exhausted. If another source of the compound is not available, the compound must be removed from the library.

When a library stored in microplates is first made, compounds are located in specific wells of a master plate. When the library is resolubilized, it is generally not feasible to solubilize the compounds in the same order and place the compounds in the same well position of the master plate. Even if this were attempted, any depleted compounds would result in empty wells in the master plate. For these reasons, most compound management organizations do not attempt to get the same well location. It is known that every compound available is in the library, but the exact location on a plate is irrelevant. All that is important is that the location is known so screening results can be associated with compound ID.

Whatever the criteria for defining a library, it is generally not a good idea to add additional compounds to the library that meet the criteria. Additional compounds can be added to a supplemental library with a different name. This is important so researchers have mechanisms of knowing which compounds have been evaluated in a screen and which have not. At a minimum, the supplemental library needs to be differentiated by a date.

4.2.4.2 Custom Plates

Custom plates contain specific requested compounds in a specific format. They are requested for a specific purpose. This contrasts with plates coming from the plate replication system that are specifically for high-throughput screening.

There are many reasons for requesting custom plates; thus, the custom plate system must have considerable flexibility. Some common reasons for requesting custom plates include:

- Small number of compounds needed for secondary assay.
- Confirm initial hits from an HTS assay.
- Measure dose–response curves for confirmed hits from an HTS assay.
- Evaluate newly synthesized or purchased compounds in a primary assay.
- Deconvolute multicompound wells from a multiplexed assay.
- Screen small library in a specific assay.
- Screen list of compounds obtained from computer search of the entire collection looking for compounds similar to hits in a primary assay—may be based on substructure searches or other properties.

4.2.4.2.1 *Source of Compounds.* The most common source of compounds for custom plates is solubilized samples set aside for this purpose. Microtube and plate storage systems were described in the solubilization section of this chapter. These systems are used to store single-use microtubes, multiuse microtubes (multisipping systems), or copies of master plates reserved for custom plate purposes. (Opening of master plates to make custom plates is not recommended because of the problems associated with exposure to moist air.)

Another source used in many companies is the source plates used in primary screening. Some screening organizations believe the best way to confirm hits is to use another aliquot of compound from the same plate used to prepare the assay-ready plate. This approach confirms that the screening result is repeatable, which is important in understanding the repeatability of the screening assay. In this approach, the dose–response evaluation will uncover if there is a problem with the compound handling leading up to the screen.

In the latter case, the replication process must produce screening source plates with sufficient volume to allow an aliquot to be obtained for confirmation purposes. The screening process must also save and protect screening plates so that the confirmations can be performed on unchanged compounds after all the data have been evaluated. This approach also requires a specialized liquid handling system to perform the "cherry picking" process. This is usually performed outside the compound management area.

4.2.4.2.2 *Sample Plate Formats.* Every researcher has a favorite plate format. A major challenge is trying to pare down the number of formats that must be supported and still meet the realistic customer needs. Sometimes the decision is to only support certain variations and let the customer reformat the plates.

Even the basic screening format can be up for dispute. For the basic 96-well plate, are all wells used for compounds or are some wells reserved for controls and blanks? Where are the controls and blanks located? It is common that a company will have a standard plate format that is used for all HTS assays and also for most other assays. This degree of standardization is important for the ease of programming robotics and also to reduce complexity in computer systems and calculations. For the basic 96-well plate, it is common to leave 1 or 2 columns of 8 wells blank, with the remaining 80 or 88 wells available for compounds. Others like to spread the empty wells around to study edge effects in sophisticated assays.

The 384-well and 1536-well plates are easiest to deal with if they are treated as multiples of the 96-well plate they are based on. Thus, if an 80-well format is used for the 96-well plate, then a 320-well format is used for the 384-well plate. Some feel the use of 64 wells for controls and blanks is too many.

Additional variables to deal with are compound concentration, solvent variations, and volume. With primary screening there are generally a couple of concentrations and solvent blends favored. These are generally carried over to custom plates. Specialized assays may require a greater range of compound concentrations. Some secondary assays require more compound; thus, volume in the custom plate may be important.

Dilution series can also be represented across a plate where one or more compounds are in the individual wells but the wells vary in concentration across the plate. The substantial variety in the plate approach enables great flexibility in screening approaches. For dose–response studies, the layout of each compound on the plate becomes important; generally, there is a top concentration and successive dilutions of this concentration in adjacent wells on the plate. Variables are how many dilutions are needed, what is the degree of dilution in each successive well, and which direction on the plate do the dilutions occur.

The complexity of these issues can be illustrated in the following list of formats supported by a recent custom plate system:

- 80-well basic format, columns 1 and 12 empty
- 1:2 dilution series, 4 wells per compound
- 1:2 dilution series, 8 wells per compound
- 1:4 dilution series, 4 wells per compound
- 1:10 dilution series, 4 wells per compound
- ∫ log dilution series, 4 wells per compound
- ∫ log dilution series, 8 wells per compound

In addition to these basic plate layouts, each dilution series format was offered with the series duplicated on the plate. Also, each plate was offered in four different top concentrations and in either 100% DMSO or 30% DMSO. All plates had the same volume.

Another custom plate system in the same company did not offer the dilution series layouts, but offered a wider range of concentrations and a variety of volumes.

4.2.4.2.3 Custom Plate Automation Strategies. The automation strategy for custom plate systems tends to be to get as much money as possible and spend it wisely. This is the area of compound management that has the most impact on the largest number of users. Plus, automation of this area has the double impact of improving throughput and service factors while also improving quality. In addition, the custom plate systems can provide capabilities that most researchers would not have dreamed of asking for before the system was available.

Fully integrated systems start with a massive storage system to store millions of microtubes, combined with a sophisticated robotic picking system to allow the withdrawal and return of several thousand tubes daily, integrated with liquid handling work cells capable of withdrawing liquid from the racks of microtubes and manufacturing custom plates according to the specifications of the customer. All this is possible without operator intervention except taking completed plates off the stackers and delivering them to the customer.

Fully integrated systems are available from a handful of vendors/systems integrators, and each one is custom built to the specifications of the company. Service, support, initial cost, time to implement, and the space required are major considerations. Fully integrated systems housing large collections typically cost $1 to $5 million. Alternative systems described below vary widely in cost but typically fall in the range of $50,000 to $500,000.

If the budget or management support is not available for a fully integrated system, a number of alternatives are possible. The microtube store can be smaller and not directly integrated to the liquid handling work cells. The work cells can be less automated with less flexibility on plate formats and can use operators to feed the liquid handlers and seal and label plates.

Plate or rack-based storage systems offer a clear alternative. The storage systems can be manual or semiautomated. The withdrawal and return of plates or racks can be manual. The major drawback of plate-based systems is the fact that every compound on the plate is exposed to humidity every time any compound on that plate is requested. Plate-based systems tend to be relatively slow, limiting throughput and increasing humidity exposure. With the availability of piercable microtube closures, a microtube rack-based system can overcome the humidity exposure problem and deliver high-quality custom plates. Not all liquid handlers can pierce these closures, so options are somewhat limited.

In deciding whether to put in a fully integrated system or a system of lesser capability, in addition to the budgetary considerations, other factors need to be evaluated:

- What are the needs of the organization? What throughput requirements are anticipated? If the needs are modest and the maximum throughput is a few hundred compounds per day, then a manual system should be satisfactory. If large requests or deconvolution of multiplexed assays is anticipated, resulting in daily throughputs of several thousand compounds, then a large, automated store is needed.

- What are the support capabilities of the organization? A large, complex system has many sources of failure. If the organization or the vendor/system integrator cannot safely support the system, then simpler alternatives should be evaluated.

General-purpose custom plate systems are generally customized by the equipment vendor to meet specific needs. Additionally, internal automation groups are likely to play a hand in building custom plate systems. Most of the modifications from standard liquid handlers involve programming changes and additional software modules and interfaces.

4.2.5 Handling of Biologicals and Other Substances

Up until now, it was assumed that the sample collection of interest was comprised of small molecules. This is often the case where larger more complex entities are of interest in drug discovery testing. Some of the most successful drugs on the market are antibodies/proteins, which require a manufacturing and handling process that differs substantially from the small-molecule process. Similarly, the study of natural products often requires the handling of biological broths, extracts, and mixtures.

In the case of single purified proteins, it must be noted that these entities differ from small molecules primarily in size and complexity of energetics. The biological activity of proteins is based on the folding of the peptide chains into their active configurations. The activity of proteins can be damaged by heat, light, physical shearing, and chemical contamination. To prevent the loss of biological activity, which could be misinterpreted as a lack of biological activity in a bioassay, special consideration must be given to the storage and handling of proteins to retain their active structure. The specifics of these processes are very dependent on the protein of interest and are best left to the suggestions of the protein chemists that are supplying the compounds.

The case of natural product mixtures or broths carries many of the same complexities as proteins, along with the added complexity of being sample pools. When dealing with mixtures, the desired path is often to screen the mixed samples for biological activity and then go back to any active samples and determine which specific component in the mixture is biologically active. This deconvolution process is often complex as the biological activity may have been caused by one or more interacting components in the mixture. In general, the rules of handling mixtures are best determined by those that have prepared the mixtures.

4.3 Cross-Process Support Activities

The following sections describe some nonproduction activities of the compound management organization. These activities typically establish a framework in which the production activities can meet customer needs. Failure to adequately address these areas can lead to general failure of the compound management organization. In some companies, achievements in these areas are the primary basis for judging the compound management organization.

4.3.1 Quality Control

Quality control is meant to address the daily, routine activities carried out by all staff to ensure they are performing their activities as expected and are delivering a quality product. Included in this are start-up procedures with every piece of equipment to confirm that the equipment is working correctly that day. Quality control is carried out by operators.

Every production procedure should have specific start-up and shutdown steps as part of the documentation. Checking that liquid handlers deliver solutions precisely and within specification should be part of the daily system suitability assessment. Also included in the daily assessment should be a visible and physical check of the layout of each equipment deck that the proper modules are in place, tubing is secure and pliable, solutions are fresh, and so forth. There should be a daily confirmation of dispense accuracy and precision and an evaluation of carryover.

4.3.2 Quality Assurance

Quality assurance is an activity of management, supervisors, and engineers to build robustness into processes. Part of having robust processes is designing into the process the various QC checks necessary to ensure the process is operating correctly. It is a management responsibility to make sure the QC procedures are developed, documented, and carried out daily.

Another part of quality assurance is building the knowledge and understanding of how products, labware, solutions, and the like perform over time. There should be ongoing studies evaluating container seals, solvent quality, decomposition of solutions, effects of storage conditions, and so forth. Seals and containers should not contaminate compounds or solutions with traces of plasticizers or other additives. There should be an ongoing process developed to evaluate incoming solvents and labware for contamination and performance.

Evaluation of customer complaints is probably best performed in a QA area. In this manner, knowledge of the causes of problems can be used to develop more robust processes.

4.3.3 Training

Training provides special challenges in the compound management area because of wide differences in scientific understanding, experience, and abilities of the staff. Production activities should be robust processes that do not require a high degree of technical understanding. Operators need to be able to follow the procedures with special attention on what to look for to identify problems and when to call for help.

The compound management supervisory staff generally has a good understanding of the fundamentals of the processes. The major challenge is building an understanding of how the output of compound management is used in discovery and providing leadership skills suitable for managing a largely nontechnical production staff. Building enough understanding of computer systems, automation, and robotics to communicate effectively with engineering staff is important.

A special challenge in training for compound management operators is dealing with high turnover and modest technical skills. Many compound management organizations make considerable use of temporary workers. These are typically not long-term employees, and a training program must be devised to get this caliber of worker trained to perform quality work in a minimum of time.

4.3.4 Process Engineering

There is a need for ongoing activities process engineering and process improvement to better understand existing processes and to develop improved processes and equipment. There should be an emphasis on evaluating new technology such as seals, denser plate formats, nanodispensing, and improvements in liquid handling equipment.

Part of this effort is staying current with information technology advances as they impact discovery. New informatics concepts will impact the compound management area eventually, and it is best not to be surprised.

Another part of this effort is staying current with popular concepts in drug discovery, such as high-throughput screening, parallel synthesis, and profiling compounds for safety and metabolism properties. One important approach to staying abreast of industry directions is attending scientific conferences and trade shows. This is a mechanism of maintaining an awareness of what the competition is doing and also technical advances of the vendors and systems integrators.

4.3.5 Bar Coding

Bar coding is an important concept in compound management and requires some separate discussion. The conventional linear bar code symbology (similar to the bar codes seen in grocery stores) and equipment used are not too important. This is an established technology that works well. Codes 39 and 128 are two symbologies that are commonly used. Symbologies are standardized formats for the bars and spaces between them that are agreed upon in industry. These two symbologies are convenient for laboratory work because of the number of characters usually encoded and the physical size of labels that can be attached to containers. Laser scanners should be used wherever possible because of their accuracy and ease of use.

Conventional linear bar codes are good for labeling most containers and locations. Two-dimensional bar codes are used on microtubes because the high-density information matrix allows full identification in the small area available on the bottom of a tube and because readers are available that allow the simultaneous reading of the bar codes on 96 tubes in a rack at one time. Two-dimensional bar codes could be used for all containers, but the cost of readers does not make this practical.

Different organizations use bar codes in different ways. Some companies attempt to build intelligence into the number, with certain digits signifying certain information. This leads to problems as exceptions are made and processes change. The authors have found best results when the bar code contains no information. The bar code needs to be nothing more than a unique identifier. There is no need for any information on the bar code label other than container ID. Related information is easily stored in a database.

Two-dimensional bar codes are relatively new to compound management and equipment/labware vendors. Two-dimensional bar codes are a misnomer, since there are no bars involved. The two-dimensional bar code is a collection of dots with a specific arrangement that allows the placing of a great deal of information in a small area, typically within a circular area of a few-millimeter diameter. This format is small enough to be placed on the bottoms of microtubes and other small containers. The specification allows for optional human-readable characters around the circumference of the two-dimensional bar code, but a magnifying glass is required to read these characters.

A number of problems have occurred in recent years with two-dimensional bar codes, and readers are cautioned to be careful with the details of this technology and to perform substantial testing and due diligence at all phases of the project. Do not assume final products will perform like prototypes. The technology for applying two-dimensional bar codes to microtubes is evolving and there is a lack of industry standards. It is easy to get into a situation where production changes result in a reader not being able to read all tubes in a tube store. Obviously, this is a situation to be avoided.

Bar code labels for vials and plates must be robust. They must withstand DMSO, humidity condensation, and cold temperatures. For vials where the tare weight is important, it is critical that the label not change weight over time. It should not absorb water, and the adhesive should not give off any volatiles. Typical labels use special papers or plastics combined with an adhesive. The materials must be carefully evaluated for chemical resistance and weight stability.

Recently, some companies have been using new processes to apply bar codes to glass vials, eliminating the need for paper and adhesives. These processes involve etching the glass or applying ceramic coatings. Although this technology is promising, printing equipment is usually limited to specific suppliers and the contrast is sometimes not sufficient for the use of existing bar code readers.

An interesting technology that may have some potential in compound management is radio frequency identification (RFID). In RFID, an RF tag can be affixed to a container such as a vial or plate, much as a bar code would, and the sample information in the RF tag can be read from some distance using a transceiver, not necessarily in line of site. This could be useful in finding missing containers within a lab or storage location. The disadvantages are higher cost and difficulty in limiting a reading to a single container. Thus, there would be difficulty in determining which container was being used if multiple containers were on a robot. To be used effectively, it would be necessary to use both bar codes and RF tags on a container. This technology is commonly used to identify animals in animal toxicology studies, stockyards, wildlife releases, or pet control when RF tags are embedded in their ear or shoulder and the animal can be positively identified from a distance without contact, but we do not know if any compound management organizations are using RFID technology to manage containers.

4.3.6 Customer Service and Logistics

A compound management organization and the individuals in it are generally judged and evaluated primarily on customer service. Compound management organizations can usually benefit from specific personnel dedicated to the customer service function.

One function of customer service is to establish and monitor service-level agreements. Customers should have a realistic expectation of when services will be performed. There should be an internal watchdog to monitor whether these expectations are met. Frequent contact with customers can help an organization focus its priorities and avoid misunderstandings.

Many customer service failures of compound management organizations occur after the main production activities have been completed. Getting the promised materials delivered to the customer in a timely manner is very difficult. Many logistics functions are ideally embedded in a customer service function. In this way, customer service provides an internal service to the compound management organization by removing barriers to providing timely service. Packaging items for shipping is not something every operator needs to know how to do. Likewise, keeping an adequate inventory of shipping supplies and other consumables is something customer service can have as a responsibility.

Customer service should be the initial contact with most customers regarding questions and complaints. However, customer service has a delicate task in that it must defer some questions to the production organization or management. Even here, however, customer service can make sure questions are answered and followed up on.

Customer service is also a good place to put some miscellaneous tasks, such as controlled substance management, compliance with hazardous materials requirements, purchasing, financial, and performance reporting, samples for outside investigators, and other administrative tasks that frequently get bungled if let to a scientific type to perform.

In defining customer service responsibilities, it is important to understand the line to be drawn between systems and people and to worry about how much automation is needed. Many systems can be simplified at considerable savings of programming effort if a person can be used to manage some difficult issues and interfaces. For instance, if there are multiple inventory sites, a person can search for a compound in various locations, rather than write complex computer code to find the compound, and make a rational decision where best to get it from. There are times when the best design decision may be to put a person in the middle of an automated process.

Part of the customer service logistics is taking responsibility for shipping to other sites, whether they be international or down the road. How best to support multiple locations is an issue for many compound management organizations, and the customer service group is a good place to maintain this focus. Customer service is also the best place to deal with special procedures for shipping solubilized compounds. Often the decision is to ship at dry ice temperature, and this requires special packing, handling, and paperwork.

Production operators usually worry about robotics, efficient handling of compounds, and protection from contamination. But when it comes to large, complex orders, such as thousands of plates, or splitting production orders to ship to multiple sites, it is easy to get confused and ship the wrong stuff to the wrong location. The final QC of orders to make sure they are complete and shipped to the right place with the right paperwork at the right time is an ideal responsibility for a customer service function.

The customer service organization can also help greatly with internal efficiency by assisting with internal delivery and movements when portions of the compound management organization are not ideally situated. It is common for production areas to be split between different laboratories and buildings. Customers are easily confused on where to deliver new compounds and where to pick up their orders. Customer service can provide a single open doorway to the compound management organization. It can facilitate production operations by moving finished orders from the production area to a centralized receiving and distribution area.

4.3.7 Strategic Planning

The compound management organization should regularly create a strategic plan with significant input from its customers. Experience indicates that compound management activities are cyclical in nature, and part of the strategic planning effort is deciding what is important at the current time and what can effectively be deferred. For example, a small collection may indicate that a major effort on compound acquisition is needed. Or, a solubilized library can last many years, but at some point a major effort is needed to resolubilize it.

The strategic planning effort should highlight:

- Evolving and new customer needs.
- Areas for process, technology, or automation improvement.

- Areas for information technology improvement.
- Needs for compound acquisition or replenishment.
- Benchmarking against the competition.
- Efforts needed to stay competent.
- Budgeting and return on investment (ROI), specifically discussion of maintenance and lack of importance on ROI.
- How to remove bottlenecks already in place.
- Focus on not creating new bottlenecks.

4.4 Information Technology Needs

Information technology is essential to the success of the discovery organization. A well-integrated system promotes efficiency and effectiveness of the organization at each step of the process. Successful implementation of a high-throughput screening approach to drug discovery requires sophisticated information technology support because of both the complexity and sheer volume of data involved. Systems should be integrated to support sharing of information and optimized rational processes. Systems should be modular to allow changes and enhancements in key areas while keeping the overall process functioning.

In many regards, the e-commerce model employed by companies such as Amazon.com fits the needs in the compound management area. Customer focus is maintained by providing tools to make ordering compounds efficient and effective. Powerful, multisite and multiorganization inventory data are summarized to let the customer know what is available and how soon. Once ordered, powerful tracking features are employed to keep the customer informed of the status of the order. Order fulfillment can come from a number of different distribution and manufacturing sites. A widespread logistical system allows partial orders from many directions to be assembled into a fulfilled customer order in minimum time.

Following this simplified, high-level description, it is useful to look at a general systems architecture that supports this model, review the major features required of each component of the integrated system, explore how the integrated system is used to support key aspects of the discovery process, and, finally, explore some concepts of system design that make it possible to maintain the system and keep it up to date with emerging requirements.

4.4.1 Generalized Systems Architecture: Research Execution System

The research execution system is an integrated collection of applications that support all aspects of the discovery process. Proper integration enables the execution of the discovery process with the most effectiveness, speed, and efficiency. The architecture diagram in Figure 4.2 depicts the major applications and information flows required for such an integrated system.

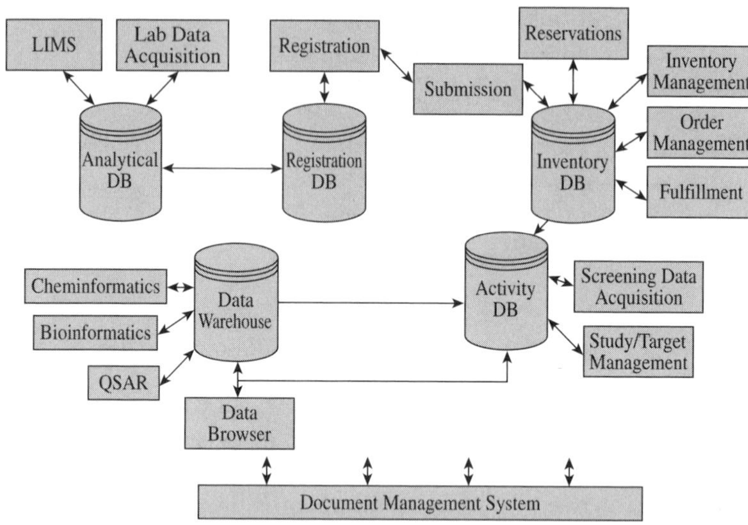

Figure 4.2 Example of the research execution system architecture. (LIMS: Laboratory Information Management Systems, QSAR: quantitative structure-activity relationship.)

Although the discovery organization needs access to numerous applications, the compound management activity only requires a few to do their work. In the following sections, most emphasis is placed on those applications and data directly supporting the compound management activities. The other applications are briefly described only to set the context for the overall application architecture.

It is important to keep the user community in mind when reviewing this diagram. The user community is made up of chemists, therapeutic area biologists, screeners, cheminformatics and bioinformatics staff, and compound management personnel. Each category of user has different needs and interests.

4.4.2 Requirements for Compound Management Applications

The following applications are directly involved in compound management activities and are described in some detail. It is important to recognize that for several of these applications the primary users are not compound management personnel, but the customers of the compound management organization.

4.4.2.1 Registration, Analytical, and Chemical Properties

When a new compound is obtained, through either purchase or synthesis, the chemist must determine the compound's structure, purity, and some chemical properties before the compound is submitted to biologists for activity screening.

Using structure drawing tools, the chemist inputs the structure into the registration system. This system then evaluates the structure for reasonableness and determines whether the compound has been registered previously. If the structure has not been registered previously, a new compound number is assigned to the structure. If the structure is already used, the registration system must assign the previous compound number, but a new batch number to uniquely identify the compound.

Registration systems are typically linked to other databases that store analytical and chemical property data. This information can be entered at the time of registration or later as additional data are obtained.

Registration systems may also be involved in the submission process to compound management. Submission is the beginning of the inventory process. It is important to recognize that submission does not always occur at the time of registration. Submission of the compound to the compound management organization may occur at the time of registration or at any time thereafter. Many companies do not require that all compounds be submitted. It is common to provide a container ID and amount submitted as part of the registration process.

Registration systems may also be directly involved in the requesting and reservation processes. If the chemist submitting a compound knows that a portion should be sent quickly to a biologist for testing, it is common to allow this request to be made at the time of registration. In addition, it is common to allow the chemist to reserve all or a portion of the submitted amount for specific scientists or purposes.

4.4.2.2 Submission

Submission is the process of delivering some or all of a compound to compound management for inventory and customer requests. Key information required for submission is container ID (a bar code number), compound identifier (compound and batch numbers), and amount. The amount may be a weight or, for solutions, a concentration and volume. Submission results in creating an inventory record in the inventory management system.

The submission process is also a common place to stipulate storage and handling conditions and to identify hazards.

4.4.2.3 Inventory Management

When a compound is submitted, an inventory record is created. Because compounds are sometimes lost or forgotten, it is a good idea to have a receive step in compound management that time-stamps the database record to indicate the compound has actually been received into inventory. At this point, compound management may perform a process to verify the amount received.

The first step in inventory management is to create the initial record for the container and to indicate it has been received. Most commonly, the second step is to put the container away and to record a storage location in the database.

The inventory management system should hold transaction records that identify each change in inventory for the contents of a container. For a neat dispense, the date of dispense, the amount dispensed, the dispense container ID, the dispense request ID, the requester, and the dispenser are typically recorded. The dispense process must update the inventory amount remaining.

Inventory management systems should be capable of managing both neat compounds and solutions. They should be able to handle different types of containers, such as tubes, vials, and plates. For a plate, they should be able to produce a plate map describing the contents of each well in the plate. The well is just another container type and should be capable of holding either dry compound or solutions.

A big problem in compound management is maintaining inventory accuracy. The inventory management system should support various processes for cycle counting that result in inventory records being updated. A common process is to periodically use a bar code scanner to verify the storage location of every container. Another process is to perform a check-weigh of a container to verify the amount contained. This may require the system to maintain separate tare weights and gross weights for each container.

The inventory management system database is commonly the place where all submission and reservation information is stored. In addition, all dispense, movement, and current inventory transactions should be stored. Information about pending orders and completed orders may also be stored in the inventory database.

The inventory system should include query and reporting functions necessary to support compound management activities. Examples include:

- Generate a plate map.
- Report everything known about a container, including transactions and existing orders.
- Display all containers with a given compound or lot.
- Display current location of a container.

4.4.2.4 Reservations

Two types of reservations must be supported. The first is user reservations. User reservations are made by the compound owner and designate how much of a compound is reserved for various users or purposes. The user reservation is typically made by the chemist submitting the compound. The user reservation will typically have an expiration date so the compound can be made freely available to others after the reservation expires. It is important to have reservation management functionality, including notification prior to reservation expiration, canceling of reservations, editing of reservations, creating a new reservation, and renewal or extension of reservations.

System reservations are made by the order management system and are maintained by the inventory management system. When an order is placed, the system should look at the amount available, and if the amount requested is less than the amount available and meets other rules, the system should place a reservation for the amount requested. At the time of dispense, the system reservation is removed and the inventory amount is decremented by the amount actually dispensed. In this way, it is not possible to order more than is available and inventory accuracy is maintained.

Philosophically, it is important to have a robust reservation system in order to build trust in the compound management organization's ability to provide stewardship to the inventory. Without an effective reservation system, there is no assurance a compound will be used for its intended purpose. Without this assurance, it is common for the chemist to simply refuse to entrust compounds to compound management.

Some systems include a management report identifying compounds below a specified inventory level so that inventory replenishment actions can be taken. Sometimes this is limited to specific compounds of continuing interest.

4.4.2.5 Requesting and Order Management

Some companies choose to make a distinction between requests and orders. A request is what a user has asked for. The request is converted to an order after an approval mechanism is executed. The approval may involve a check against business rules or approval by management. If multiple sites and containers are involved, the request conversion to order may require the reservations be placed against specific containers at specific sites. In other cases, the approval is automatic and the difference between a request and order is largely semantic, depending on whether the request has been written to the database and system reservations placed.

Versatility and ease of use are key to successful implementation of request and order management systems. The user interface is critical. Different customers have different needs, and it is very difficult to meet everyone's needs. The customer may be a biologist looking for a couple of neat compounds, or another biologist looking for 10,000 specific compounds on custom plates, or a screener looking for a library of 500,000 compounds. A system optimized for one type of request may be very unpleasant to use for another request. Combine this with the changing needs and tools of discovery and it is easy to understand how the order type may not have been understood or known at the time of system design.

Considerable effort must be made in gathering requirements for the requesting system. If the system is to be built rather than purchased, prototyping is an excellent way to refine the requirements and do a better job of meeting needs. The request module is also a place where system performance must be considered in the design. It is not unusual to request several thousand compounds in a single order, and the user will not be happy with slow screen-to-screen movements and slow processing. It is important to be able to import and process large lists of compounds. Users will want lists of accepted compounds and rejected compounds so they can attempt different routes of obtaining needed compounds.

Once the request is made and converted to an order, the user wants feedback on the status of the order, including an estimate of when it will be delivered. Functionality to modify and cancel orders is needed.

The ordering system must be cognizant of what the compound management organization can actually make. This is particularly important for custom plates, especially if more than one custom plate system is available and different systems offer different compounds, concentrations, or formats.

4.4.2.6 Fulfillment

Fulfillment systems are only for the use of compound management personnel. These systems receive the order information for each order to be assembled at a given site and provide the needed information to the automated systems in use at that site. The fulfillment system is the integration point between complex automated systems, such as automated storage systems, custom plate systems, or replication systems, and the ordering and inventory management systems.

In some cases, it is acceptable to build the fulfillment system as part of the inventory management system. For large organizations, with multiple sites, particularly if there is concern for reliability of communications between sites, it is often desirable to create individual fulfillment systems at each site. In this case, if the main inventory management system fails, work at the individual site can continue until communications between sites have been restored.

4.4.2.7 Activity Data and Screening Systems

Activity data and screening systems are not directly used by compound management personnel, but typically must be interfaced to support some specific actions.

When library plates are delivered to a screening laboratory, plate map data describing what compounds are in each well of a plate with a specific bar code number are usually transferred to the activity data system used by the screening laboratory. If activity is associated with a well of a plate, the screening laboratory can then identify the compound associated with that activity. The scientists can then evaluate activity levels without depending on the inventory system.

When compounds with potentially interesting activity are identified, additional amounts of the compounds are usually required for further evaluation. It is sometimes effective to order these compounds through a direct interface that can transfer a list of active compounds from the screening system to the order management system.

4.4.2.8 Other Applications

Several other applications are typically part of the research execution system. Compund management personnel usually have no need to access these systems. Brief descriptions of these applications are included to complete the context of the research execution system architecture.

4.4.2.9 Laboratory Information Management System

A Laboratory Information Management System (LIMS) is often used to manage analytical activities and store analytical information.

4.4.2.10 Instrument Data Acquisition

Instrument data acquisition systems are often used to interface to sophisticated analytical instruments to assist in the collection of raw data from the instruments and to calculate meaningful results from the raw data. Data acquisition systems are often interfaced to a LIMS to transfer results information into the LIMS database.

4.4.2.11 Electronic Laboratory Notebooks

Another level of data recording used in some companies is electronic laboratory notebooks. These systems typically replace paper notebooks with the intent to improve regulatory compliance, reduce efforts, and improve the ability to search for information.

4.4.2.12 Cheminformatics and Bioinformatics Applications

Many companies use a variety of cheminformatics and bioinformatics application programs to calculate data and perform searches. These applications may need access to registration and chemical property data to identify interesting compounds. They may also need access to the inventory system to determine which compounds are available for ordering.

4.4.2.13 Structure-Activity Relationships

Structure-activity relationship (SAR) or quantitative SAR (QSAR) applications are often used to understand compound activity and to decide what compounds to attempt to synthesize for further studies. A SAR application needs access to compound ID and structure information from the registration system.

4.4.2.14 Toxicology and Metabolism

Toxicology and metabolism systems are LIMS for storing and interpreting animal study and cellular data results.

4.4.2.15 Document Management

Document management systems are used to create, store, retrieve, and query documents. Many documents must have strict version control procedures in place.

4.5 Organizational Structures

Figure 4.3 shows an example of an organization chart for a compound management organization.

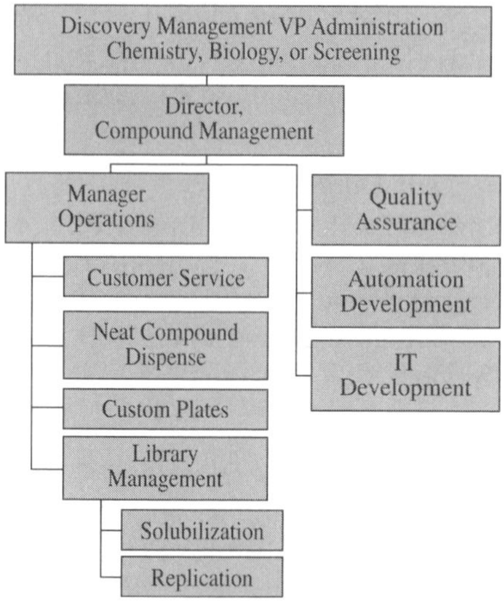

Figure 4.3 Possible compound management organization structure.

Compound management can fit in various areas of the discovery organization. It is often part of chemistry or research administration. It can be part of IT, screening, or biology—in any of these, management needs to pay special attention that all needs of the research organization are met, not just those of the area to which it reports. Avoid conflicts of priorities.

4.6 Barriers and Obstacles

Various barriers and obstacles exist to implementing an effective compound management organization. The emphasis should be on achieving success through increasing flexibility and reducing complexity.

4.6.1 Caution on Technology Investments

Expenditures on automation technology should focus on places where improved technology will lead to significant improvements in processes. Top priority should be placed on delivering new capability, better quality, or faster service.

Information technology needs to be carefully designed so it can have a significant lifetime. Software should be modular with clean interfaces that can be modified without making major changes to underlying code.

Automation and IT should be flexible, with multiple inputs and outputs possible. Hard automation will often not meet long-term needs. It is difficult to anticipate future needs and changes to processes, but if an automation and IT is implemented with various input and output points, it is likely to meet future needs as processes evolve.

Carefully evaluate benefits of complexity and move toward simplification. Do not automate too much. Avoid large expenditures in areas like vial picking and placing where manual labor is cheap, automation is expensive, and manual processes provide sufficient quality.

4.6.2 *Rate of Change in Biological Testing Processes: Dynamic Nature of Sample Formats*

HTS assays and secondary screening processes and formats are constantly evolving to employ improved technology and decreasing amounts of compounds. This is good, but it does introduce the need for added flexibility in the compound management organization. Changing sample formats is often the rate-limiting step in bringing new services into use.

4.6.3 *Process Standardization*

Every company has a different way of executing key processes. The costs of automation and speed of implementing new techniques can be optimized if maximum use is made of industry standards developed on a formal or ad hoc basis by Society for Biomolecular Screening (SBS), Laboratory Robotics Interest Group (LRIG), and Lab Automation. Such standardization makes it possible for vendors to deliver products more economically and quickly while improving reliability.

References

1. I Yates. Compound management comes of age. *Drug Discov World* 35–42, Spring 2003.
2. GW Kuroki. Automating parallel organic synthesis. *Am Lab* 32:49–51, 2000.
3. B Halstead. *DMSO Handbook: A Complete Guide to the History and Use of DMSO*. Cancer Book House, 1981.
4. Representative Web sites with scientific papers and vendor listings: www.labrobotics.org, www.labautomation.org, www.miptec.com, www.sbsonline.org.

chapter 5

High-Throughput Screening

Mark Beggs, Victoria Emerick, and Richard Archer

Contents

5.1 The Development of High-Throughput Screening within Pharma

5.1.1 History and Drivers

Pharmaceutical research is classically divided into four phases, commencing with target identification and ending with preclinical. Target identification (TI) is the process by which the hypothesis of the involvement of a particular molecular target is postulated. The next stage, lead identification (LI), hopefully delivers several chemical lead series that show a demonstrable effect on the disease target of interest. Today, high-throughput screening (HTS) is an integral component of the LI process. Successful LI programs result in a project's transition to lead optimization (LO), where medicinal chemistry is brought to bear on the optimization of the structure-activity relationships (SARs) around specific pharmacophore classes that were identified in LI. Finally, optimized lead compounds from LO programs enter preclinical, where their overall selectivity and toxicity profiles are assessed as a precursor for their entry into clinical development. Each of these steps is associated with a significant attrition rate for reasons of lack of efficacy, specificity, or toxicity (1) (Figure 5.1).

Screening has been an integral part of the pharmaceutical industry from its earliest stirrings. Indeed, the identification of the antisyphilitic Salvarsan by Paul Ehrlich and Sahachiro Hata in 1910 involved the screening of some 900 compounds in mice and bacterial assay systems (2). At the time, there was little knowledge of the molecular nature of the interaction—only the empirical observations of efficacy and selectivity. As understanding of the molecular nature of disease improved, it became possible to target specific enzymes and receptors, leading to the position where, once the disease hypothesis had been

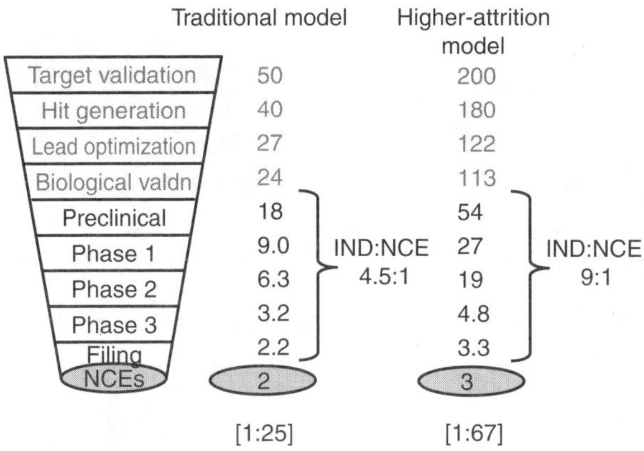

Figure 5.1 Typical attrition rates across pharma R&D, from early research to new chemical entity (NCE) approval. Two models are illustrated: the traditional model where discovery programs commenced with a validated target, and the recent higher-attrition model accounting for the impact of larger numbers of unvalidated targets arising from genomics initiatives. (Data from Lehman Brothers internal analyses (1) with modifications.)

postulated, it became routine to screen diverse compounds against the target in an entirely empirical approach. During the golden age of the pharmaceutical industry (1950 to 1990), empirical screening came to the fore as a pragmatic approach to dealing with the issue of a target-rich environment, yet one that had little associated understanding of the molecular basis of the disease. The approach has worked well, with important classes of drug discovery through empirical screening. In an analysis conducted for the U.S. National Bureau of Economic Research, 17 of 21 drugs (80%) cited as "having had the most impact on therapeutic practice" are shown to have been derived from empirical screening approaches (3) (Table 5.1). These encompass almost all the major disease sectors in which the pharmaceutical industry has operated. The remaining four drugs (20%) were identified through rational discovery approaches.

Over several decades of their existence, pharmaceutical companies have amassed considerable collections of compounds and natural products. These have arisen not only from in-house drug discovery programs, but also from unrelated activities, including chemical and molecular dyes (ICI, Novartis, Roche), fine chemicals (Merck), petrochemicals (Sanofi-Synthelabo), agrochemicals (ICI, Dupont, Novartis), and explosives (Dupont, ICI), whose available chemistry skill sets and legacy compound collections formed the basis for the company's nascent pharmaceutical programs.

Historically, *in vitro* screening was regarded as very much a service function for the chemistry teams. It was required to respond to requests to screen specific chemical series or "diversity collections" present within the existing compound collections. The screening process was characterized as an entirely manual operation in which one or two full-time equivalents

Table 5.1 Sources of Drugs Having the Most Impact upon Therapeutic Practice (Introduced between 1965 and 1992)

Generic Name	Trade Name	Indication	Date of Enabling Discovery	Date of Synthesis	Manufacturer	Lag from Enabling Discovery to Market Launch
Classical: Discovered through Empirical Screening of Compounds in *In Vivo* or *In Vitro* Pharmacology Screening						
Cyclosporine	Sandimmune	Immune suppression	NA	1972	Sandoz	
Fluconazole	Diflucan	Antifungal	1978	1982	Pfizer	7
Foscarnet	Foscavir	CMV infection	1924	1968	Astra USA	67
Gemfibrozil	Lopid	Hyperlipidemia	1962	1968	Parke Davis	19
Ketoconazole	Nizoral	Antifungal	1965	1977?	Janssen	16
Nifedipine	Procardia	Hypertension	1969	1971	Pfizer	12
Tamoxifen	Nolvadex	Ovarian cancer	1971	NA	Astra Zeneca	21
Mechanism Driven: Screening against Known or Suspected Disease Target						
AZT	Retrovir	HIV	Contentious	1963	Glaxo Wellcome	
Captopril	Capoten	Hypertension	1965	1977	Squibb	16
Cimetidine	Tagamet	Peptic ulcer	1948	1975	SmithKline Beecham	29
Finasteride	Proscar	BPH	1974	1986	Merck	18
Fluoxetine	Prozac	Depression	1957	1970	Eli Lilly	30
Lovastatin	Mevacor	Hyperlipdemia	1959	1980	Merck	28
Omeprazole	Prilosec	Peptic ulcers	1978	1982	Astra Zeneca	11
Ondansetron	Zofran	Nausea	1957	1983	GlaxoSmithKline	34
Propranolol	Inderol	Hypertension	1948	1964	Astra Zeneca	19
Sumatriptan	Imitrex	Migraine	1957	1988	Glaxo Wellcome	35
Drugs Discovered through Fundamental Science						
Acyclovir	Zovirax	Herpes	?	?	Glaxo Wellcome	
Cisplatin	Platinol	Cancer	1965	1967	BMS	13
Erythropoietin	Epogen	Anemia	1950	1985	Amgen	39
Interferon beta	Betaseron	Cancer, others	1950	Various	Berlex	

Note: Shown are the relative contributions of empirical and rational drug discovery to the effective management of human health over a 28-year period (1965–1992). Analysis conducted for the U.S. National Bureau of Economic Research. (CMV: cytomegolovirus; BMS: Bristol-Myers Squibb; BPH: benign prostatic hyperplasia.)

Source: Adapted from IM Cockburn, RM Henderson, in NBER Innovation policy and the economy. AB Jaffe, J Lerner, S Stern, Eds., MIT Press, Cambridge, MA, 1:1–34, 2001.

(FTEs) tested, at most, a few hundred compounds per week against an individual target. The logistics, both within the screen itself and in the surrounding processes, such as compound supply, were poorly considered and informatics was almost nonexistent (4). During the lifetime of a typical project, its screen would test 5,000 to 20,000 compounds over the course of several years. At its conclusion, the project would have screened only 5 to 20% of the available compound collection.

Toward the end of the 1980s several screening groups across the industry developed and "sold" the concept of HTS internally within their companies. The belief was that through internalizing the emerging capabilities in assay technology, instrumentation, automation, and cheap available computing power, significant enhancements could be made in their capacity to screen compounds.

Initially progress was slow, but during the last decade of the 20th century, claimed throughputs increased by at least two orders of magnitude to a sustainable 10,000 to 40,000 compounds per week per target (4,5) as shown in Figure 5.2. Groups screened their compound collections as either singles or mixtures of known compounds, depending on individual philosophy.

The drivers for such a major overhaul in capabilities were severalfold. First was the belief that it was desirable to screen the entire compound file rather than a subset whose selection was based on human intuition. Indeed, one of the basic tenets behind empirical screening is that it would lead to the identification of hitherto unknown structure-activity relationships. Second, the development of HTS-friendly homogeneous assay technologies, including, among others, scintillation proximity assay (6) and time-resolved fluorescence spectroscopy techniques (7,8), obviated the need for complex screen protocols and greatly improved throughput rates and data quality. Third, the availability

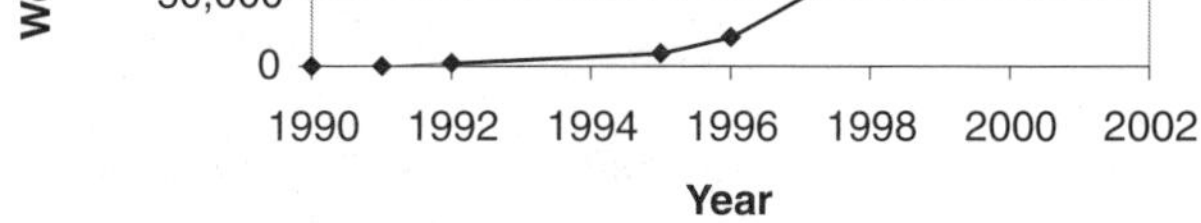

Figure 5.2 Historical increases in point screening capability for HTS from 1990 to 2001. Based on improvements up to the 2001 performance level, a 1-million-compound file could be screened in 4 weeks. Data are compiled from authors' own sources and represent industry realized capacity for individual HTS screens on a per campaign basis. In practice, each screen usually operates once a week with a time base that allows for up-front preparation, screening, compound management, and data analysis activities.

of microplate-based instrumentation, associated automation, and computing infrastructure supported the integration of various HTS elements to the point that data capture, compound information, and quality assurance/quality control (QA/QC) procedures could be handled relatively seamlessly.

This rapid expansion in screening capabilities raised within the industry a hitherto entirely academic question: Is it most efficient to screen the entire compound collection or focus on screening subsets of the collection driven by chemical intuition?

5.1.2 *Semirational vs. Empirical Screening Approaches*

The received wisdom during the early 1990s was that an established HTS screen operated in a semirational, as opposed to empirical, mode in support of the chemistry team. Typically, pharma assembled validation and diversity subsets of its compound collection, comprising a few thousand compounds at most. The balance of the compound collection was often open to the project's chemists to make their screening selections. The screening group would receive and test the chosen compounds on a screen-specific basis. The resulting compound "hits" would be reported to the chemists, who would initiate analogue searches of the compound collection, which in turn would be submitted for screening. Multiple iterative cycles would be undertaken before the screen was deemed completed. In practice, the screen, and its accompanying FTE resource, was retained for the duration of the LI/LO project, being required to screen the output of ongoing new chemical synthesis. This process would typically run for 3 to 4 years.

A holistic consideration of the process timelines led to the conclusion that iterative screening–analysis–compound ordering cycles were inefficient when lined up against the rapidly rising capabilities of the HTS groups. If it were possible to screen the entire compound collection over 6 to 8 months, why spend 18 months on iterative rounds of screening?

Second, the general improvement in compound management capabilities allowed the timely provision of screen-ready sets of compounds in microplate format. This capability was most effective when the compound collection was organized into a standard microplate-based library, or set of libraries, which could be quickly replicated, as opposed to the project-specific reassortment of compounds required for iterative rounds of screening. The cost to the organization of repeat iterations is high in terms of tracking, management, and sample preparation (9). Third, the available HTS resource could not sustain supporting multiple projects with expected project life spans of 4 to 5 years. In order to support the wider objectives of the organization, it was necessary to complete the screening campaign in a relatively short timescale and move the resources onto the next LI campaign.

In general, it is the case that the major pharma companies have moved to a position where they screen their entire compound collection against a disease target. For reasons of increasing the probability of earlier success, and subsequent accelerated transition to a LO project, compound collections have, in some cases, been prearranged into multiple nonequivalent target class-specific collections. This in itself does not impact the efficiency of the process, and the logistics can be supported by a modern automated compound store. This approach lends itself to scenarios where pharmas operate a systems or target class approach to discovery. It is obvious that several discretely organized sets of the compound collection can be created and maintained in support of classical target classes, including seven transmembrane receptor, protease, nuclear hormone receptor, kinase, and ion channel targets.

This position notwithstanding, assuming a screen-ready target, leading-edge HTS facilities are capable of screening the entire corporate compound collection within a timescale of several months. Given this capability, the savings realized by segmenting the compound collection into target-specific collections is at best only a month or so.

The authors are aware of organizations still retaining the semirational approach to screening. The reasons cited include financial savings in reagent/consumable usage and the perceived high specificity of the organization's compound collection. In practice, neither of these arguments withstand close scrutiny.

Reagent provision is essentially a fixed-cost business, and costs do not scale linearly. This is in contradistinction to the belief held at the laboratory level that cost increases linearly with scale of operation. In practice, substantial discounts can be obtained for predictable bulk orders, and vendors are open to engaging in discussions along these lines.

A basic tenet of empirical screening is that it can result in the discovery of hitherto unknown structure-activity relationships. Until recently, hard evidence that the value of HTS was dependent on the size of the compound collection screened was difficult to substantiate. One recently reported meta-analysis conducted by SmithKline Beecham's HTS group tracked the impact of increasing the number of compounds screened against both the number and potency of the declared hits (Figure 5.3). Over a 5-year period the number of compounds screened increased 40-fold. This was accompanied by a modest (4.6-fold) increase in the number of leads but, importantly, a 30% increase in the overall lead potency (IC_{50} values decrease from 5.4 to 0.1 μM). To our knowledge, this analysis is the first public disclosure that argues strongly for whole as opposed to semirational library screening. With this data in mind, it is hard to see a valid intellectual reason against screening the entire available compound collection.

Approaches to subscreen the compound collection based on structural similarity have been criticized on the basis of insufficient predictability and resource drain (9). We believe that arguments currently advanced against empirical screening strategies reflect either an absence of key infrastructure capacity or the difficulty of controlling the influence of the organization's combinatorial and medicinal chemists. Some pharma organizations report increased project timelines as a result of moving away from whole library screening toward a focused library approach.

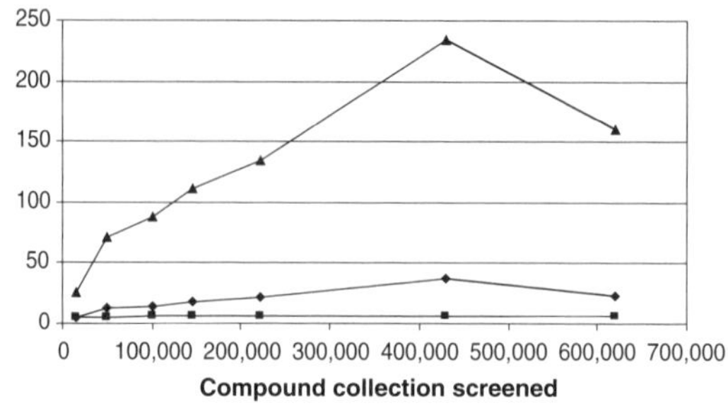

Figure 5.3 Impact of increasing file size on HTS hit potency and frequency. Data from an analysis of HTS campaigns conducted by the Screening Sciences Department at SmithKline Beecham. Shown are the frequency and mean potency of hits arising from HTS campaigns that screened compound collections of increasing sizes. Squares represent average potency of HTS hits; diamonds represent number of HTS leads arising; triangles represent the product of mean potency and lead number. (Data are courtesy of Dr. T.D. Meek, GlaxoSmithKline.)

5.2 High-Throughput Screening Today

5.2.1 Point or Sustainable Capabilities

One of the great disconnects within the HTS discipline today is the fascination with point performance, as opposed to sustainable metrics. Indeed, the casual observer at a screening conference could easily leave with the impression that every screening facility is replete with automation platforms running screens at throughputs in excess of 100,000 compounds per day in 1536-well microplate formats. Over the last few years, many presentations have been made concerning the ability of a particular assay technology or automation platform to operate at the level of 100,000 to 200,000 wells screened per day (10). It is understandable that HTS groups wish to promote their capabilities and justify the significant expenditure in enabling capabilities. However, in terms of measuring how close the pharma organization is to achieving the business objectives for research, point productivity metrics are not particularly useful.

5.2.2 Realized and Aspirational Performance Metrics for HTS

Obtaining a "fix" on the delivered output of a pharma's HTS function is a useful diagnostic of the overall capacity of its discovery organization. It is not surprising that the realized output falls well short of the installed capacity within the screening facilities. The HTS process represents the point of convergence for a number of up-front activities upon which HTS is critically dependent for its end-of-year performance metrics. The two key metrics are the number of campaigns mounted and the integral number of compound wells screened per annum.

Our own examination of the year-end output (2001 to 2002) of a top 10 pharma's HTS annual output reveals that some 40 to 65 LI campaigns are typically mounted against a compound file size of 300,000 to 400,000. These numbers are in broad agreement with the metaanalysis recently conducted by Hi-Tech Business Decisions (11) (average compound file screened 442,000; 3.3 HTS sites per pharma organization and 20 LI campaigns per site).

It is instructive to examine the utilization of the HTS facility operating at this level. A large pharma HTS facility therefore delivers some 12 to 26 million screening results per annum. In practice, this is currently achieved by screening mixtures of 5 to 10 compounds assembled into common microplate wells, thereby reducing the actual number of wells screened 5- to 10-fold. Consequently, an annual HTS output of 7 to 14 million screening data points is not atypical for a typical pharma organization (assuming a 50:50 representation of singles and mixtures screening). To deliver this output, 8 to 12 screening lines are typically established within the organization. Based on the above scenario, calculation of the mean screening throughput of the installed screening systems reveals daily throughputs of between 2300 and 7400 results per system. If we assume a capacity of 100,000 results per system per day, this represents only 2 to 7% utilization of the installed capacity. Senior discovery managers privately agree with these observations, as they match their own experiences when visiting their HTS facilities unannounced.

The utilization argument should not be interpreted as a criticism of HTS facilities or their management. Indeed, their critical dependence upon upstream business processes, most notably assay development, protein/cell production, and compound management activities, suggests that HTS output is a very useful diagnostic for the wider discovery organization. To date, excess screening capacity has been established, and indeed has been necessary, due to the high level of unpredictability and uncertainty behind the delivery of HTS-ready assays to the screening groups. The current lack of effective coordination of LI processes requires the HTS facility to operate in a high-response, low-efficiency mode analogous to the city fire department. For example, a fire department that operates on a high-efficiency, low-response basis requires maintaining a high forward workload, which would not be particularly useful to society. If more proactive management of resources can be brought to bear, we can see HTS being able to move to a high-efficiency, lower-capacity model. The argument as to how this might be achieved will be developed later in Section 5.4.3.

With respect to near- to mid-term aspirations for overall LI capabilities, there is a surprising level of agreement across the larger pharma companies. Organization goals are in the region of 60 to 125 targets per year, with a desire to screen approximately 1 million compounds per screen. This throughput is 4- to 10-fold in excess of the current metrics (Table 5.2). Achieving a 4- to 10-fold increase in HTS output impacts most of the early discovery

Table 5.2 Aspirational and Delivered HTS Capabilities

	Aspirational Targets	Realized Today (2003)
Compound collection screened	1 million	~400,000
Campaigns completed pa[a]	60–125	25–50
Request cycle time (ex RA)[c]	5–6 months	9–15 months
Overall HTS cycle time	3–5 weeks	19–39 weeks
Assay adaptation cycle time[b]	1–2 weeks	3–9 weeks
Primary HTS cycle time[b]	1–2 weeks	7–12 weeks
Confirmation/IC_{50} cycle time[b]	1 week	9–18 weeks
Enhanced output package	Potency, selectivity, cellular target, physiochemical parameters	Little past potency determination

Note: Shown are currently achieved and aspirational metrics for HTS/LI.

[a] Assumes steady-state conditions where number of campaigns completing pa equals those starting. Typically this workload is spread across two or three screening sites.

[b] Allows for interprocess time in advance of work starting.

[c] Reflects the elapsed time from request for screening services to the completion of the output package.

(Data from authors' own compilations.)

research organization to a similar extent, and significant improvements in productivity will have to be made across a broad front, notwithstanding the current climate of fixed research headcount levels.

5.2.3 *The Cost of High-Throughput Screening*

Given the considerable investment into HTS and associated disciplines over the last decade, it is reasonable to question the overall return on investment. The 2001 estimates for the size of the global HTS market were in the region of $1.7 billion (11). This figure reflects annual expenditure by the industry in terms of capital, consumable items, and technology access license fees. Internal costs, including labor, facilities, and services, are excluded. Assuming this expenditure is spread over approximately 400 HTS laboratories worldwide, this approximates to an annual expenditure in the region of $3 to $5 million per laboratory (11). In practice, the real cost to the pharma organization of its HTS operations is much greater than the visible cost of consumable and capital expenditure. Most research organizations do not track the true cost of supporting LI campaigns, regarding FTEs and building and facilities costs as part of the given cost of maintaining a research capability.

The authors have conducted their own analysis of the cost to pharma of mounting an LI campaign. The scope of activities includes assay development, protein production, and operational screening from primary screening through to potency determinations. We have applied current industry norms (2001) for the mixture of assay technologies employed, the impact of screening mixtures of compounds, a detailed breakdown of reagent costs, and

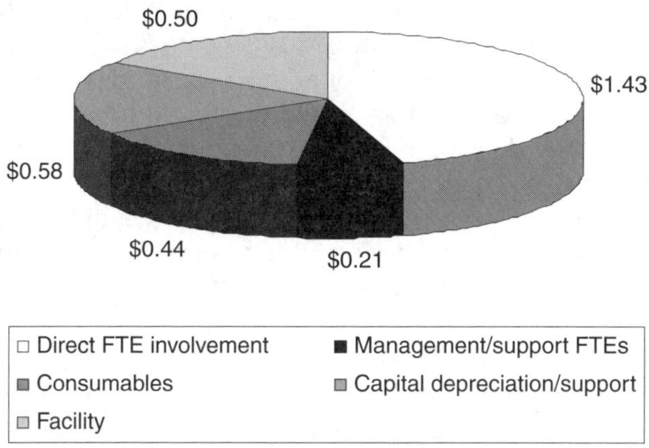

Figure 5.4 Breakdown of the cost of HTS. The authors' analysis of the full cost of mounting an HTS campaign is summarized. Inputs to the model were typical industry percentages for screen types and reagent costs. Activities in scope include assay adaptation, protein expression, and HTS. Total employment costs (direct and indirect cost elements) are estimated at $116,000 per FTE. Management and support FTE costs were estimated as 15% of direct FTE costs. It is assumed that each HTS campaign screens 400,000 compounds and the facility mounts 25 LI campaigns per year.

known facility and building costs. Our analysis suggests the true cost to the organization for a given target is in the region of $3.15 per compound screened. Of this total, consumable and reagent expenditure represents only 13% of the overall cost; labor and facility costs contribute the greater proportion of the total (Figure 5.4).

In consequence, assuming the typical large pharma prosecutes some 40 to 65 HTS campaigns per annum and screens an average of 400,000 compounds per screen (Section 5.2.2) at a cost of $3.17 per compound screened, the total cost to the pharma organization of this undertaking is between $50.7 and $82.4 million per annum.

The current attrition rate model suggests a 1:67 survival rate for early-stage research projects achieving new drug application (NDA) approval (1). On this basis, our derived $82.4 million figure represents approximately 10% of the total cost of bringing a new chemical entity (NCE) to market (assuming an integral cost to market of $802 million) (12). Given that these derived costs for LI represent a significant proportion of the total research and development cost, it is not unreasonable to ask about the success to date of HTS.

5.2.4 High-Throughput Screening Successes to Date

Attempting to analyze high-throughput screening's overall impact on research and development is not simple for three reasons. First, the long R&D pipeline times mandate that few compounds, identified as a result of HTS activities over the last decade, will have reached the market. Second, the starting point of HTS is hard to define. Clearly, overall capabilities have advanced significantly since the late 1980s (Section 5.1.1, Figure 5.1); a sustainable throughput of 10,000 compounds per screen per week a few years ago would at best be regarded as a medium-throughput capability today. Third, proving that a development compound did not have its conceptual roots in a lead series originating from HTS activities is difficult owing to the understandable desire of the wider organization to claim leads for its own. This latter factor will undoubtedly result in underestimates of the successes of HTS. The concatenation of these factors complicates the answer to an obvious question.

5.2.4.1 *Direct Impact of HTS on Clinical Trials and Beyond*

The Hightec Business Decisions meta-analysis (11,13) identified some 62 HTS leads, over the period 1985 to 2000, that have been subsequently tested in man. Assuming 5 to 10 years' time elapses between HTS testing and phase I trial commencement, this suggests that the 62 leads were identified over a 6- to 11-year period across at least the 23 major pharma companies participating in the study. Assuming steady-state conditions throughout, this output translates to 0.25 to 0.45 clinical trials per pharma company per year having HTS origins. There is clearly a major difference between this level of output and the number of clinical trials initiated per annum around novel chemophores. The time frame for the above analysis is wide and encompasses a three-order-of-magnitude increase in HTS capability over the period 1985 to 2000. The authors believe that the HTS contribution to clinical testing is underrepresented in the analysis. Over the period 1985 to 2000 a significant amount of LI activity was maintained within therapeutic area teams, often running in parallel with nascent HTS programs. It is possible that many HTS leads were subsumed into therapeutic area team activities and their origins have been subsequently obscured within organizational memory. Clearly, rational design and external licensing programs will have made finite contributions to clinical trial initiations over this period; however, given the relative contributions of rational and empirical approaches to health care (Table 5.1), it is hard to attribute the missing balance of clinical inputs to anything other than empirical programs, irrespective of their ownership within the pharma organization.

The same study cited the 2001 total output of lead compounds identified as 326 across 44 HTS groups (23 major pharma companies listed). Assuming a best-case (1:1) relationship between leads and target and a 1:41 LI-to-NCE survival rate (1), this should yield a total of eight NCE approvals from 2001 screening activities. Assuming the leads segregate evenly across the 23 major pharma companies (and disregarding biotech contributions), this predicts that an annual yield of 0.3 NCEs per company will result from current HTS activities. Clearly, this level of output is an order of magnitude below the current predictions for sustained pharma profitability (14,15) and provides further justification for the need to increase both the number of HTS campaigns mounted and the number of compounds screened (Section 5.2.2).

Viramune (nevirapine) is the one publicly cited example of a currently marketed drug that was discovered as a result of HTS. Boehringer identified a lead compound in 1985 after screening 1600 compounds against an HIV reverse transcriptase target. Development of the lead compound led to the synthesis of nevirapine that was ultimately approved for use in 1996 (16).

There is much discussion in the industry that "HTS doesn't work." In practice, it is true that there is insufficient output from the LI phase to provide an adequate flow of NCEs at the other end of the pipeline, assuming current attrition rates. However, this is not a philosophical weakness. The real issue is that if HTS groups are only delivering approximately 20 to 50% of their required rate of data generation, then it should not be surprising that the downstream output is correspondingly diminished. Put another way, if HTS was operating at the required throughput levels, it would meet the business objectives set for it. The reasons for this deficit are primarily cultural or organizational and are not due to limitations in technology. The drug discovery factory concept is attempting to redress this position (see Section 5.4).

5.2.4.2 Indirect Benefits of HTS

The development of HTS as a discipline has positively impacted several other areas in pharma outside clinical trials.

The ability to rapidly screen large numbers of compounds has allowed HTS techniques to be employed as a means of identifying "tool compounds" for developmental pharmacology. In particular, generic assay technologies have been developed to exploit HTS capabilities in the hunt for orphan receptor ligands (17).

The desire to increase throughput has resulted in the development of more facile and predictable automation platforms that can be deployed and commissioned in relatively short periods without significant integration or workup costs on the part of the pharma organization. Such stable platforms no longer require the pharma organization to support major in-house automation projects with long development times and uncertain outcomes; instead, it can concentrate its efforts on core activities.

Assay miniaturization has largely been driven by HTS activities. Although probably initiated by the prospect of reduced consumable costs (see discussion in Section 5.1.2), this initiative has significantly improved the industry's capabilities. During the mid-1990s, almost all screening was performed in 96-well microplates in reaction volumes of >100 µl. The current position is that most screens employ a 384-well format and reaction volumes of less than 30 µl. Further initiatives to move some technologies to 1536-well format and reaction volumes of only a few microliters are ongoing and will become the norm for specific applications. These advances represent considerable improvements in signal detection instrumentation, liquid-handling capabilities, and homogeneous assay technology and reflect the combined efforts of multiple HTS teams and their technology providers. Many of these technologies have subsequently been adopted by TI and LO research groups.

5.3 Requirements for HTS

In order to establish and sustain an HTS capability, it is necessary to address not only issues of screening facility infrastructure, but also those of informatics, compound collection, cell culture, and assay development capability. These are discussed in the following section with particular reference to issues currently confronting the screening manager.

5.3.1 Compound Collections and Infrastructure

5.3.1.1 Library Access and Design

The size and diversity of the corporate compound collection is regarded by many pharma companies as one of their few unique assets in the discovery field. At the time of this writing, the current paradigm is to screen the entire available compound collection against targets of interest, preferably as "singles." The tendency is to move toward screening a collection of at least 1 million compounds, and a general feeling still pervades that the more compounds that can be screened, the better. While this empirical antiintellectual approach still rides uncomfortably within the context of the discovery organization, it reflects the aspirational goals of most organizations.

Internal debate is ongoing as to whether the increasing extent of chem-informatic and physiochemical profiling data can be used to weed out all compounds that do not possess drug-like qualities. This approach would avoid wasting time on HTS leads that would never survive LO. Although this strategy is becoming practicable, the current view is that such a strategy is undesirable as the SAR information derived from even a "no hoper" compound may still provide a valuable input to an SAR program.

Most companies have consolidated their compound collections in a single store where compounds are stored as dry powders. This represents the corporate archive and significant effort is expended to ensure that all new syntheses are added to the collection. With advances in chemical synthesis techniques, the absolute mass of compound delivered each synthesis is decreasing. As a consequence, companies increasingly supply the needs of all early discovery screening processes from solution stores in a drive to conserve the amount of each sample within the collection. Such solution stores are often maintained in close proximity to the major points of consumption using a federated architecture. Significant investment has been made over the last decade in modern automated sample stores to maintain and manage both dry and solution stores. This technology has been built on automated warehousing principles employing sophisticated tracking software and process hardware (18,19). Such stores have automated the entire manual process beginning with sample registration, through storage and retrieval, to preparation and supply of assay plates to the relevant screening facility, while providing full sample identification through use of sophisticated process and environmental control with full audit capability. At a macrolevel, this technology can be regarded as mature and stable. However, the output formats from compound stores continue to evolve with increasing interest shown in ever-lower volume and contactless liquid dispensing approaches.

The details of the specific storage conditions for compounds have long proved a subject for inconclusive debate within pharma. Most of the early storage conditions were defined with little recourse to substantiating data as regards the stability of compounds stored within. The variables include temperature, oxidizing atmosphere, humidity, solvent vehicle, and freeze–thaw cycling. Recently, pharma has started to investigate the impact of these variables on the stability of compounds maintained for long periods (20–22). Although properly outside the scope of this chapter, some early conclusions are emerging. The most significant factor that appears to affect stability is humidity control of dimethyl sulfoxide (DMSO) stored samples. The impact of freeze–thawing cycles remains open (21,22).

Some companies have undertaken significant collection characterization exercises to determine the true representation of their compound collections. Organizations are understandably reluctant to share the findings of such studies; however, where published, the observation that 30 to 40% of the compounds do not possess the registered structure or the necessary purity required is not uncommon. It is likely that more rigorous structural integrity and purity checking for newly synthesized compounds will become the norm.

While the provision of automated storage capabilities has become a relatively commoditized market, the major differentiator lies with the compound management software outside simple machine control software. Pharma organizations report considerable frustration with their attempts to integrate compound management on a multisite or global basis. The sophistication of the organization's true requirements is often overlooked at the time of hardware purchase. In practice, relatively sophisticated software architecture is needed to support any scientist within the organization being able to view available stock and location of any compound within the organization, as well as place orders to the stores against a managed set of business rules. Furthermore, algorithms need to be defined to support routine replenishments, from multiple points of input, and top-up activities across the federated collections.

5.3.1.2 Library Composition and Formats

Historically HTS campaigns would encompass both medicinal chemistry and natural product collections. Recent trends show a move away from natural product screening, at least as an in-house activity. While the demonstrable utility of natural products as a source of pharmaceutical diversity is not at issue, the reality appears to be that the elapsed time and organizational cost associated with the deconvolution of natural product mixtures lies outside the current expectations for the LI process. Those organizations still undertaking natural product screening usually prefractionate extracts in advance of screening as a means of reducing the effort requirements downstream.

Much debate and activity has ensued around the wisdom of screening compounds as preassembled and therefore known mixtures or as single entities. Early attempts to combine several compounds into a well resulted in confusion over the identity of the active entity and the presence of artifactual results. More sophisticated approaches evolved whereby two-dimensional arrays of mixtures were generated (23). This proved more successful and typically resulted in a four- to fivefold reduction in the number of wells needing to be screened. Although popular for a time, this strategy appears to be falling out of favor in most large-size organizations. The precise reasons are not clear, although the availability of high-capacity screening systems as well as a more holistic consideration of the overall compound management and screening process may be arguing against its application (24).

5.3.2 Cell Production Requirements

Assay formats for HTS that utilize whole cells provide three advantages over classical biochemical formats. First, they obviate the requirement for high-level expression and purification of target proteins. Second, they provide the means to evaluate the interaction of the target with test compounds in an environment that resembles its native environment. Third, they provide the means to select active compounds that are limited to those that can traverse cellular membranes to access their intended targets. Clearly, this last condition can work against the identification of novel SAR elements, which are selected against for reasons of cellular accessibility yet, in themselves, may drive SAR programs.

Given recent advancements in cell culture methodology, recombinant techniques, and platform technology for detecting intracellular molecular interactions, it is expected that cell-based assay formats will continue to increase within the HTS portfolio. Additionally, many cell-based HTS assay formats involve growing numbers of test wells. Increasing the number of test wells for cell-based screens or the number of cell-based screens has created new operational and scientific challenges.

Notwithstanding their stated advantages over classical biochemical format assays, cell-based assays impose additional requirements on the HTS laboratory. The first is for the consistent and timely provision of high-quality cells as a key input to the screen. Experience with the operation of cell-based screens has shown that interbatch variations in cellular response parameters have to be controlled carefully. In several cases variation has been traced to differences in the cell-handling procedures performed by different human operators. This is understandable as the tolerances and criticalities of the cell processing operations are not well defined. The second issue is one of coordination of supply and consumption. The growth cycle time of many cell lines imposes a 2-day lead time before HTS may commence. Under the working practices current in pharma discovery, this has limited the days when cell-based screens can run to the second half of the week.

This position clearly impacts the overall throughput of cell-based screens, especially in cases where the batch run size imposes additional capacity limitations.

The technology for fully automated cell culture was first established in the late 1980s. However, until recently, its application was primarily within the biopharmaceutical and vaccine manufacturing sector (25,26). Over the past 5 years, this automation has been successfully applied to HTS reagent provision (27,28), but it has typically been limited to the provision of large volumes of intracellular proteins and membrane fragments, rather than the delivery of live cells. Today, process automation designed specifically to meet the needs of the cell-based HTS environment has been established and is becoming an essential technology in many organizations. Experience with the use of this technology has demonstrated improvements in both the throughput and the interbatch consistency of cellular HTS assays. It has proved applicable to the wide range of cell lines investigated to date (29,30). Initial concerns that automation would not be able to perform manipulations with the dexterity of a human operator have not been substantiated, and good equivalence of cellular response data is being demonstrated (30). One unexpected organizational benefit resulting from such initiatives is that cell culture conditions can be reliably and rapidly recreated within different parts of the pharma organization as the culture conditions and manipulations are fully defined and articulated within the robot.

Although cell culture automation now supports the up-front preparation of assay-ready cells on an out-of-hours basis, many organizations have yet to engage with the coordination issues required to optimally use both cell culture and HTS resource. Some groups are deploying handheld Palm Pilot-based technologies as a means of enabling bidirectional communication between cell culture technicians and a central process database (30). This clearly represents a step in the right direction, but a wider scoped application covering the multiple groups involved in bringing HTS campaigns to successful conclusion is still required to optimize tasking (see Section 5.4.3) and avoid unnecessary waste.

5.3.3 HTS Infrastructure

The early years of HTS saw significant investment by organizations in anthropomorphic single-armed systems feeding microplates to multiple workstations responsible for reagent addition, incubation, plate washing, and signal detection. Indeed, the ownership of such systems was regarded as a badge of office. The overall throughput of such systems was predicated by the transaction time of the anthropomorphic arm and, as such, was considerably lower than that which could be maintained by a human operative. Arguments were advanced that the automation could work out of hours; however, this rarely withstood examination as the reliability of the peripherals and reagent stability issues precluded extensive out-of-hours work.

This period resulted in the appearance of two distinct camps of organization: those who designed their assays around simple-to-run processes and ran them manually using simple workstations, and those who relied on anthropomorphic armed systems and their accompanying team of support engineers and system integrators. Experience showed that the net benefit to the organization was greater in the former case. Groups experienced in HTS operation reported that batches of up to 200 to 300 microplates (16,000 to 32,000 compounds in 96-well format) could be screened on a sustainable basis using a workstation approach (31). This was approximately four times what was being achieved using the automation of the day. The underlying reason for success was twofold: first, the screening and informatic methods were optimized for operation on a batched basis; second, careful consideration of appropriate staff profiles ensured that screening staff were not demotivated by a routine task.

The 300-plates-per-day batch size probably represents the upper limit to what can be attained, on a sustainable basis, through the workstation approach (assuming single-shift operation). In practice, detailed examination of intraplate assay parameters revealed significant variation as a result of batch-wise processing. While the performance parameters lay within acceptable limits, demands for better data quality from the workstation users, as well as increased throughput from the users of anthropomorphic systems, resulted in the development of a new generation of ultra-high-throughput screening (uHTS) automation systems. Such systems employed the principle of multiple parallel processing of microplates such that the overall productivity of the system was not limited by the transport capacity of the earlier systems (32–34). As a result, throughputs in excess of 200,000 compounds per day per screen have become achievable. As such, the throughput capacity of modern uHTS automation is probably fit for purpose; it is now possible to screen a one-million-compound library as "singles" within a working week. It should be noted that these throughputs are currently unsustainable, week in and week out, and wider issues of workflow coordination are required to deliver this throughput on a sustainable basis. These are discussed in Section 5.4.

Improvements will no doubt continue in the areas of low-volume liquid handling as well as reader sensitivity and throughput. The major issue confronting purchasers of uHTS automation is one of mean time between failures. Traditional laboratory workstations and robotics were never designed to operate on a high utilization basis, and as such, component reliability is far from the desired position. The development of uHTS factories (see Section 5.4) where throughputs of 200,000 compounds per day per screening system are required will result in single HTS runs consuming $30,000 to $40,000 of reagents alone. As such, system uptime, reliability, and vendor support packages will become increasingly important elements in the purchasing decision. In the extreme, pharma organizations may choose rental options or complex result-driven payment terms, rather than outright purchase, as a means of ensuring adequate system reliability.

5.3.4 Informatics Architecture

5.3.4.1 What Is Needed

On first inspection, the core informatics requirement associated with HTS appears relatively straightforward. Each screen is supplied with a set of microplates containing test compounds. The microplate sets have been created from a master set of plates whose compound locations and identities are known. Each microplate set carries a unique set of plate identifiers, allowing their genealogy and contents to be indexed from the master set. Each screening campaign divides for reasons of capacity into a series of batch runs of typically 100 to 280 microplates. It is usual for an individual screen batch to be processed on a daily basis.

During the course of the screen protocol, the assay signal is measured on a specified instrument capable of measuring the specific signal produced during the assay. Signal types include radioactivity-induced photons, fluorescent light intensity, anisotropy and polarization, luminescence, and absorbance photometry. The instrumentation generates a data value for each microplate well measured on an individual plate basis. At the conclusion of the screen the data analyst merges the compound information with the assay signal data sets and performs a simple normalization algorithm that expresses the effect of each compound in terms of internal control wells usually located within each microplate:

$$\text{Percentage effect of compound} = \frac{(\text{Compound assay signal}) - (\text{Screen blank signal})}{(\text{Screen control signal}) - (\text{Screen blank signal})}$$

The hardware processing requirements and data transfer transactions to support this process even at today's 100,000-throughput peak loading levels are not particularly demanding, especially when compared to other business sectors, including, for example, the financial industry.

Surrounding this relatively simple data processing lays a quality control step that is performed by the screening group at the conclusion of the screen. This process includes trend monitoring, both within a screening batch and across historical batches, as well as a number of relatively straightforward systematic error traps that allow the analyst to determine if any piece of screening hardware is performing in a substandard manner. Compound wells that are rejected by this process are rescheduled for later screening batches. Compounds categorized as active are highlighted for further screening to confirm activity and determine potency. While historically the software packages required to handle the above processes have been written in-house within pharma, increasingly they can be bought as a configurable package from one of several software houses. Although the core functionality remains constant, a significant amount of bespoke configuration is required to integrate it within the client's existing architecture.

In the view of the authors, considerably more investment is required in HTS software outside straightforward LIMS architecture if pharma organizations are to realize their aspirations for significant increases in LI productivity. One major element revolves around leveraging the productivity of the discovery organization through the deployment of collaborative planning or supply chain management principles. This is discussed in detail in Section 5.4.2.1. Collaborative planning is an approach borrowed from manufacturing industry, where it has been deployed for some time, to help manage the unpredictable element behind component supply, hardware, FTE availability, and varying product demand. Although to date the parallels between hi-tech manufacturing and LI activities have been largely unrecognized within pharma, a few organizations are starting to realize that the application of these techniques could have a major impact on overall LI productivity (35,36). Significantly, this could be realized without requiring further rounds of investment in platform technologies.

5.3.4.2 *Screen Optimization Approaches*

Notwithstanding the significant increases in overall HTS throughput capabilities resulting from improvements in process automation, informatics, and assay technology, the development of HTS protocols remains rate limiting for the process. Timescales for screen development vary depending on the nature of the target class and the experience of the group involved. In our experience, pharma groups report average screen development timelines of 6 to 9 months (see also reference 37). This compares with a total HTS cycle time (screen development to the end of HTS) in the region of 9 to 14 months (see also reference 11). Clearly a significant element of assay development is taken up with iterative rounds of cell culture/protein purification, which themselves are amenable to process automation. This notwithstanding, it is clear that this area places significant demands on available in-house skills able to develop effective and predictable screens. In practice, moving to higher numbers of screens per year can be seriously impaired by the lack of experienced assay development staff. Screen development revolves around the ability to optimize multiple experimental variables of time, concentration, pH, ionic strength, and volume. Today, the exploration of multiple parallel spaces is performed sequentially by human operatives. While the approach simplifies results interpretation, the impact is that optimization of screen protocol conditions may be suboptimal. This may have negative implications on reagent usage and overall screen sensitivity.

The concept of design of experiments (DoE) (38) has been proposed as an approach to better perform the multiple variable testing required in screen development (37). The SmithKline Beecham team (37) investigated the utility of the DoE approach using three radioligand binding and two enzyme assay protocols in a redevelopment mode. An automated workstation was employed to establish the multiple reaction conditions necessary for DoE. Of the five assays optimized, four subsequently delivered lead compounds. Significant increases in enzyme rate and reaction conditions were reported.

Further, the group reported that an experienced team comprising a DoE expert, a statistician, and a screen development scientist could complete the DoE process within 3 weeks.

Clearly, these data represent a significant improvement in screen development timescales. The wider issue will be to develop the necessary skills base within the life science teams to make the DoE tool applicable on a broad basis. From an organizational quality perspective, the application of such approaches can only serve to improve and standardize the methods by which screens are developed. Careful choice of automated screening platform able to work in a DoE mode would facilitate the assay optimization process currently required to convert a manual screening method into an automated method.

5.3.4.3 Real-Time vs. Batched Data Collection/Analysis

To date HTS data analysis has largely been performed retrospectively (see Section 5.3.4.1). The merger of microplate data and screening instrumentation data sets is frequently handled as a batch job and may, for legacy system reasons, take several hours to complete. Therefore, in many cases it is not possible for the screen operators to detect real-time processing errors and to effect any remedial corrective action. With the relatively low throughputs realized to date, the consequences of an occasional lost screen batch have not been viewed as overly critical. However, with screen batch sizes beginning to exceed 100,000 compounds, the implications of a batch failure become significant. Further, the requirement to repeat-screen an additional 10% of the compound collection may necessitate additional unscheduled rounds of protein production/cell culture, which may delay the completion of the screen. With this scenario in mind, the need for real-time QC of HTS data is becoming critical.

It may be possible to employ automated pattern recognition or trend-spotting algorithms in conjunction with established assay norms to automatically alert the screen operator that corrective action is likely to be required. Pattern recognition tools have been long available in the supermarket and banking businesses (36). There is also an additional requirement that the HTS platform's software is sufficiently flexible to allow corrective action to a committed run. Typically this might involve increasing the incubation time to generate an adequate signal.

A further consideration for real-time QC will be a potential data load issue. To date the classic HTS protocols result in the generation of a single numeric value for each microplate well. This has not imposed a significant loading on the network or processing infrastructure, as file sizes are typically in the region of 200 kb for a batch size of approximately 300 microplates. The advent of recent advances in high content screening (HCS), and the possibility that these screen formats may be employed in an HTS mode, may radically change the data-handling requirements. High content screen technologies generate high-resolution images of each microplate well, as opposed to a single numeric value. A modern HCS instrument platform generates a 2- to 3-MB image for each well processed at maximum resolution. This translates to a data load of 1.2 GB per 384-well microplate (A. Fletcher, Amersham Biosciences, personal communication, 2003). A full screening batch of 200 to 300 microplates would therefore generate 240 to 360 GB of data. This represents a million-fold increase in raw data file size.

The issues confronting the users of such technology revolve around either accepting the need to retain all raw instrument data files, thereby supporting retrospective analysis as new analysis algorithms become available, or, alternatively, learning to extract key assay parameters directly from the instrument files and discarding the original image files. The first position retains flexibility from the perspective of the scientist but moves the screening organization into the world of terabyte data farms and dedicated high-volume networks (A. Fletcher, Amersham Biosciences, personal communication, 2003). The second pragmatic position requires acceptance that the key parameters are identified during assay development/adaptation and only those data parameters specified are retained. In practice, it may be simpler to repeat the screen to measure new parameters than to invest in a significant overhaul of the site and laboratory data-handling infrastructure. This second position will require careful management of expectations.

5.3.4.4 *Using the Data*

It is generally agreed that HTS-derived data have two distinct purposes. Identification of hits against a specific target is the obvious one; however, arguably equally important is the aggregate knowledge derived from running many screens in a comparable manner. This results in a growing data set for each compound becoming available for analysis by *in silico* techniques.

Although somewhat outside the scope of this chapter, the issue of how to use the projected increases in HTS data volumes effectively is a fundamental one. Our observations suggest that from a cheminformatics perspective, there are issues with the nonequivalence of biological screening data generated to date. Concerns are expressed over the issue of unrecorded "protocol creep" during the lifetime of a screen to the extent that current corporate screening data sets on given compounds are often regarded with considerable suspicion by cheminformaticians.

Further, increased focus is being applied to issues of compound stability, especially when in solution phase. Pharma organizations are beginning to publish data on the impact of water absorption and freeze–thaw cycles on the stability of compounds used in HTS (20–22). It is likely that little information regarding the equivalence or microplate chronology of sets of screening data is available to the cheminformatician. Many of the proposed industrialization concepts outlined in this chapter will not only contribute to increased data volumes but, importantly, to improved data quality and organizational quality factors.

Looking to the future, HTS informatics will not simply focus on high-quality screen data, but must cover compound integrity and sample-handling chronologies. Three-dimensional fingerprint searches based on the target's structural information or multiple chemotypes may become the norm (5). It is possible that after the establishment and population of high-quality screening data sets, resulting from industrialized screening activities, the predictive power of virtual screening approaches may be significantly enhanced.

5.3.5 *Assay Technologies and Assay Development*

5.3.5.1 *The Role of Assay Development*

HTS campaigns begin with either the *de novo* development of a biochemical assay or the adaptation of an existing assay to an HTS platform. A biochemical assay may be described as a specific and measurable signal arising from the action of a biological agent on a relevant cellular or metabolic process. A biochemical assay is suitable for deployment in an HTS context only when its sample-to-sample reproducibility (robustness), its ratio of signal to background, and signal variability are fully optimized, and are shown to be reproducible on an automated screening platform (39). The process of optimizing a developed assay to work reproducibly on an HTS robotic platform is referred to as assay adaptation. This may require considerable reworking of any existing assay (40). This is because an HTS-based assay is performed in a fully automated mode, involving the processing of thousands of assay wells without human oversight or intervention, which differs from the entirely bench-level manual context by which the assay was developed.

5.3.5.2 *Statistical Considerations*

There are two main parameters to be considered when designing a robust biochemical assay for HTS: amplitude (and signal/background ratio) and assay variability. Both of these parameters are frequently captured by researchers in a single value, the popularly used Z′ value (39):

$$Z' = 1 - 3(sT + sB)/(xT - xB)$$

where xT is the mean of the total or high-control value, xB is the mean of the blank or low-control values, sT is the standard deviation of the total or high-control value, and sB is the standard deviation of the blank or low-control value.

It is desirable that means and standard deviations are obtained from each screening plate since the Z′ value can be used to compare the performance of the assay across the entire batch of microplates within a single screening run.

The Z′ value measures the quality of a screening assay. It measures both the signal to background and variability of the assay. In general, biochemical assays with $Z' > 0.4$ are regarded as suitable for HTS application and demonstrate an acceptably low incidence of both false negative and false positive results.

5.3.5.3 Practical Considerations

A core requirement of HTS is that it allows the rapid and efficient screening of the corporate compound collection. This has required a level of assay miniaturization above that traditionally achieved in classical biochemistry/pharmacology laboratories. There are two main drivers for miniaturization: improved throughputs through reducing the number of unit transactions required to screen the compound collection arrayed in microplates, and the conservation of valuable in-house derived reagents. This latter consideration should not be underestimated, as smaller assay volumes allow full HTS campaigns to be performed on targets where the biochemical reagents are limited, due to either poor heterologous expression or other reasons of biological scarcity.

Over the past 5 years, in conjunction with enabling developments in assay technologies and instrumentation and automation, the standard screening medium has moved from a 96-well to a 384-well microplate format. Considerable interest is expressed in the higher-density 1536-well format, but at the time of writing, 1536-format assays represent no more than 5 to 15% of a typical pharma's screen portfolio. Significant practical barriers confront the development of 1536- and higher well density assays. These include the delivery of submicroliter volumes of test compound, reagent mixing, evaporation, and high-speed reading of the assay signal. While these issues are surmountable, organizations are experiencing differing levels of success with higher-density assay formats. Our own observations are that packaged technology assay platforms currently offer a greater chance of success than in-house developed formats (41,42). Clearly, this position may change as experience grows, but currently it appears that the 384-format assay represents a pragmatic position as regards time and risk of assay development, sustainable and robust operation, reagent conservation, and assay throughput. The authors' experience to date is that there is little evidence to indicate that true cost savings or increased throughputs have resulted from 1536-format assays.

Assay miniaturization imposes its own limitations on the types of assay formats that are compatible with the required Z' values and plate-to-plate reproducibility. Assays should ideally be homogeneous, requiring no separation steps. The assay signal generated must be detectable simply and quickly. Appropriate formats include fluorescence changes, detected either by a change in molecular diffusion (or solution tumbling) or by changes in fluorescent intensity or wavelength of emission. Radioactive format assays employing proximity effects (SPA or flash plate format) remain popular for some target types, although concerns over the costs associated with managing and disposal are decreasing their application. Ion channel targets typically employ a signal transduction assay format in which intracellular calcium release is detected by a specific, fluorescent chelating agent (43). Table 5.3 shows some examples of assay formats suitable for HTS.

Table 5.3 Assessment of Suitability of Assay Formats for HTS Applications

Ideal HTS Formats	Reasonable HTS Formats	Acceptable HTS Formats
Fluorescence polarization	Scintillation proximity assay	Time-resolved fluorescence with enzyme-linked immunosorbent assay (ELISA) component
Fluorescence resonance energy transfer	Intracellular calcium mobilization (FLIPR)	ELISA
Fluorescence intensity	Reporter gene	
Luminescence		
Homogeneous time-resolved fluorescence		
Colorimetry		

Note: Shown are available screen methods by their suitability for application in an HTS role. (FLIPR: fluorometric imaging plate reader.)

Data from authors' own compilations.

From the perspective of assay sensitivity, it is preferable to develop an assay where the measured signal increases rather than decreases during the course of the assay (44). Further, assays that are continuous, whereby the signal develops over a measured time course, are preferable to end-point assays (44). In practice, however, end-point assays are highly amenable to operation in an HTS context because stacks of plates with completed assays may be stockpiled prior to reading. This avoids the issue of having to synchronize liquid-handling steps with longer cycle time signal detection steps. In this context, long read cycle times require that the assay signal remain stable for long periods. This can be for periods of up to 17 hours for a batch of 200 microplates being fed through a reader with a tick rate of 5 minutes per microplate.

5.3.5.4 *Assay Adaptation*

With few exceptions, assays developed for HTS applications have their genesis within the sponsoring TI groups. The assay undergoes its initial proof-of-concept phases operating under very different conditions and levels of scale than it will ultimately be exposed to in production mode operational screening. It is currently the case that an assay's transition from a development to an operational phase requires changes to the assay to make it compatible with the automated HTS platform and intensity of operation. This process is termed assay adaptation.

Assay adaptation is by no means straightforward and can often involve months of work and, in extreme cases, complete reconstruction of the assay protocol. Clearly, the more thought and dialogue that is put into the initial development of an assay bound for an HTS campaign, the less time that will be spent reworking. Some of the basic rules of assay adaptation include:

- Minimization of the number and volumes of reagent addition steps (three pipetting steps or fewer).
- Tolerance and stability testing of assay reagents. Operational windows of several hours at both room temperature and 4°C should be attained. Lot stability to multiple freeze–thaw cycles should be determined.
- Stability of biochemical reagents upon high dilution or after introduction of DMSO/solvents should be determined. Stability issues may be addressed by the addition of excipients that protect proteins in solutions, such as serum albumin and sucrose.
- Calibration of the assay signal: noise ratio by use of a known standard compound where available.
- Securing internal quality control standard in advance, and all relevant pharmacological/biochemical parameters (e.g., Kd, Ki, and Km values) should be determined for the biomolecule. These parameters should be compared with those arising from the original low-throughput assay format.

One issue currently confronting the industry is how to best transition from assay development to assay adaptation. A clear articulation of the hand-off point would help define appropriate standards for entry into assay adaptation. The precise pass criteria would need careful definition. However, a minimum set of entry parameters would include demonstration of an acceptable Z′ within the assay plate of choice, using the proposed screening methodology. Definition of assay controls and known pharmacological standards would further support the case for entry.

The assay adaptation team is likely to be more multidisciplinary than required for successful development. Automation specialists, infomaticians, statisticians, and logisticians will play important roles alongside the assay scientist. The collective efforts of these specialists will be required to deliver a validated assay protocol that is amenable to operation at scale and where key dependencies are fully comprehended. Many of these skills lay outside the traditional realm of the pure assay development scientist, further arguing for separation of the two disciplines.

Ultimate review and sign-off of the screen-ready adapted assay protocol should be obtained from the sponsoring client. This should involve a careful review of the key parameters investigated during the adaptation phase, together with validation data from preliminary screening runs performed using the screening platform that will be used for production screening. Included in the review package should be clear statements of approval or rejection criteria for screening data generated.

5.4 Challenges for Lead Identification

5.4.1 Drug Discovery Factories

The concept of the drug discovery factory has received wide coverage both in the pharmaceutical press and on the conference circuit in recent years. In the original paper on the subject, a deliberately emotive phrase was coined to challenge some of the then currently held perspectives (45). The term *factory* is deliberately provocative in the context of a research organization, where the perceived value and reward sets associated with good science are radically different from those embodied within a factory. This speaks to the need to address process, organization, and people issues in a holistic sense. Experience has shown that such a radical change has not been easily assimilated within pharma, and indeed, there have been many detractors within organizations.

The analyses conducted to date on the forward profitability of the pharma industry (14,15,46) are mandating significant improvements in discovery productivity. The issue facing senior executives within discovery is that of the absence of scalability of many elements. It is impractical to source, let alone recruit, enough appropriately qualified and experienced medicinal chemists and biologists. However, ways have to be found to increase overall productivity notwithstanding this position. Technology investment in isolation has not delivered the required improvements. Increasing the output of LI is now being seen as a way of catalyzing change within the wider discovery organization. The prevailing logic is that should a 6- to 10-fold increase in sustainable LI capability become realized, the upstream and downstream elements will have to increase their capabilities accordingly. Over the past 4 years pharma organizations have begun to embrace many of the concepts expounded by the drug discovery factory promoters. The first such instance was made public in 2003 with GlaxoSmithKline's (GSK) opening of a discovery factory in Tres Cantos, Spain. This facility, specified at 300,000 data points per day (47), is the first of several planned within GSK. As noted in the Evans article (47), what is significant about the GSK discovery factory is not the visible hardware elements, but rather the invisible elements required for effective planning and operational control.

5.4.2 Planning Techniques

5.4.2.1 Avoiding Technology Rounds without Business Focus

Criticism has been leveled at the pharma industry for the expensive rounds of platform technology investments made in the 1990s. These included bioinformatics, combinatorial chemistry, HTS technologies, and automation. To date there has been little evidence that these investments have resulted in any significant increase in NCEs, decreased attrition rates, or increases in compounds entering preclinical (15,48). While one interpretation of this position is that the technologies were, in themselves, somehow flawed, an alternative is that the technologies were deployed within an organizational context unprepared for them, and therefore unable to realize any benefits resulting from such technologies. There is the apocryphal story of the screening lab that screened 100,000 compounds in 1 day but took 6 months to assemble the compound sets and 3 months to analyze the resulting data points to such mismatches.

Increasingly, organizations are becoming more focused in their acquisition of technology. They require a deeper consideration of the true organizational impact of any new technology along with a clearer understanding of its value proposition. Comprehending the complex dynamics of discovery processes and workloads is complicated, and organizations are turning to techniques developed by manufacturing industry to understand these interrelationships.

5.4.2.2 Discrete Event Simulation

Discrete event simulation is a technique used to comprehend the interaction of multiple discrete activities on the delivery of an end-point product. In the context of discovery applications, this could be the multiple processes that must come together before a high-throughput screen can run. These processes include:

- The request for a screening campaign from a therapeutic area
- Assay development
- Cell culture and protein purification
- Just-in-time live cell supply
- Compound management activities
- HTS hardware availability
- Staff skills and availability
- Hardware breakdowns
- Key reagent nonavailability
- Data analysis time and decision-making steps

Process activity is frequently interspersed with interprocess gaps of varying lengths that possess their own characteristics, such as buffer size. It is recognized that many of the inputs to the screening process occur on a random and unpredictable basis. As such, the use of deterministic modeling tools, including spreadsheets, is of limited value in calculating the likely ability of an LI facility to deliver its yearly organizational objectives. Many of these processes operate as discrete tasks and are separated from each other by significant interprocess gaps, each with their own dimensions. Discrete event simulation is a software-driven approach used to comprehend the multiple process and interprocess interactions in a holistic manner. As such, it can simulate the overall impact of new hardware deployment, shift pattern change, or modify existing business processes in terms of the organization's ability to deliver its objectives. The power of the technique lies in its ability to iterate findings.

The authors have employed discrete event simulation techniques in discovery to adequately balance the installed base of hardware and shift patterns, staffing levels, and business rules against the desired annual capacity and expected portfolio representation. The outputs of such studies are used to drive capital investments with the understanding that the planned facility has been appropriately toleranced against likely real-world loading scenarios.

The additional benefit resulting from such *in silico* simulations is that they avoid the position where an inappropriately dimensioned investment has to be reworked postdelivery. In the case of automation systems, the elapsed time between problem diagnosis and redeployment can easily take 18 months, with a concomitant lost opportunity cost and organizational exposure.

5.4.3 *Operational Management*

5.4.3.1 *Collaborative Planning vs. Phone Calls*

In silico facility design will deliver facilities whose architecture is toleranced for expected workloads. Simulation does not, however, articulate how operational planning and intragroup communication is undertaken. Within the simulator it is implicit, rather than explicit, that the facility and its operatives undertake required activities in an optimal way, allowing for the unpredictability of discovery programs. Until recently, the received wisdom has been that discovery programs are so inherently unpredictable and unstable that there is little benefit in attempting to plan any screening programs until a developed screen protocol and all the key reagents and consumables have been secured. In many cases, this results in considerable delays to screening programs because a large, unprocessed forward workload is maintained in advance of the commencement of operational screening or screen validation. Are there alternatives to managing screening campaigns in this entirely reactive way?

Groups are now attempting to predict likely campaign durations through the use of Gantt charting programs. This is a start and serves as a useful tool for getting multiple groups to agree on their involvement and estimate likely durations of substream activities. With sufficient resources, it can also allow progress to be measured against the plan and serves to capture metrics. However, as a tool, it is not particularly convenient to use when working with multiple projects, each having its own specific resource requirements. Further, it is difficult to optimize the work plan across multiple projects having conflicting demands for the same resource. This becomes increasingly an issue under conditions where one project has to be rescheduled due to, for example, issues within assay development or gene expression. In practice, little is currently done to manage this other than on a reactive basis.

The manufacturing industry has struggled with the equivalent issue for many years. Ensuring the efficient use of the manufacturing facilities and its workforce with real-world perturbations of component supply or machine uptime on a short-term basis, or shifts in market demand on a medium- to long-term basis, is a core skill within the manufacturing industry. A purely reactive position is unacceptable in this world. Instead, manufacturers have accepted the unpredictability of their supply chains and markets and developed approaches to manage them.

Techniques such as supply chain management have been developed to allow manufacturers to make the best use of their resources and deliver completed products efficiently. Supply chain management is based on a series of configured software modules. Surrounding this is a significant piece of organizational change. A key element of this is devolving organizational power to junior employees involved in delivering unit tasks as part of the wider program. These teams of workers provide operational feedback on the day-to-day progress of their unit activities that form part of a wider production plan. Production planners collate the inputs not only from the work teams, but also from key component vendors and commercial groups. Scenarios are modeled to optimize delivery and groups are retasked on a dynamic basis.

Cross-industry manufacturing feedback on the impact of supply chain techniques reports improvement in service levels between organizational units up to 100%, and improved business interaction with business partners and reductions in forecast error, leading to stabilization of planning, by 20 to 30% (49). According to modeling work done by leading industry analysts, automotive manufacturers estimate a reduction of 21% on project costs and the slashing of project timelines using collaborative planning during product development (50).

Is this tool appropriate for the management of LI programs? Most of the paradigms that confront LI programs have a parallel within manufacturing. A major benefit accruing from implementing such techniques would be increased organizational transparency as regards project status and true performance metrics. To date, few discovery organizations either capture or comprehend the true time and resource costs associated with LI campaign development and operation. It is the authors' belief that the implementation of supply chain techniques will have a significant impact on overall discovery productivity (35).

5.4.3.2 *Getting the Right People*

The techniques outlined in this section are relatively unknown within the context of pharma discovery, although they are well articulated elsewhere, including in the manufacturing sector of pharma. Indeed, in the majority of cases, production management skills and techniques are neither comprehended nor valued within the current discovery organization. This is entirely understandable given the academic innovation culture currently prevalent within discovery. However, given pharma's desire to increase overall discovery productivity by 6- to 10-fold (see Section 5.4.1), some redefinition of skills and aptitudes will be necessary. Put another way, these goals are not realizable without organizational change.

To date, the process elements of LI programs are headed up by qualified life scientists. Their interest and dedication in establishing their organization's initial capability in automated compound management, bulk-scale protein production, and HTS must not be dismissed. However, this pioneering phase has now passed and there is a need for pharma to assimilate different skills sets in order to focus on delivery of organizational goals. Production-centric individuals are not normally the by-product of academia's Ph.D. and post-doctoral programs—pharma's traditional recruiting ground. Instead, pharma will have to be prepared to look to more diverse sources of skills to realize its objectives. The difficulty of this change must not be underestimated.

The industrial revolution showed that the guilds of master craftsmen, with their high entry barrier and restricted practices, were nonscalable. By analogy, the industrialization of discovery will require considerable task specialization, distributed activity, and comprehension of fit-for-purpose process tolerance. All of the above would have been an anathema to the master craftsmen of old, who fitted and fettled handmade subcomponents into a working entity notwithstanding the cost and time. Ironically, the quality of the handmade item seldom equaled that of a precision-engineered mass production product. Pharma organizations that are prepared to, and able to, recruit the necessary skill sets and proven expertise from nontraditional sources will be able to operate the production-centric elements of their LI processes on a professional and sustainable basis.

5.5 Conclusions

At the time of writing, HTS and its related disciplines have emerged into early adolescence. After a long period of development as a discipline, together with a massive development activity on the part of its informatics, automation, and assay technology components, it is fair to say that the key elements are now in place. Indeed, it has been argued that there is more than enough technology available to allow LI to deliver the needs of the wider organization. Under ideal conditions, a modern HTS organization can complete the screening of the entire corporate compound collection within a single week—something that would have been regarded as beyond comprehension even a decade ago. With this level of performance potentially available, and given the overall timelines of the drug discovery process, it is hard to justify why point screening throughputs need improve any further.

The major challenge confronting senior LI management is no longer one of technology; instead, it is one of coordination. The discipline needs to move away from repeated rounds of technology investment. Indeed, institutional investors have grown tired of hearing unsubstantiated claims as to how technology will increase pharma's productivity (51). Instead, pharma needs to address the more difficult issue of how to optimize its discovery processes. This is against a discovery culture that has historically undervalued process, delivery, and metrics.

The good news is that many of the same issues have already been addressed by other business sectors and that eminently workable and well-tested solutions exist that will prove workable in the discovery environment. The greatest challenge will undoubtedly be the cultural impact of introducing major change to both the required skill sets and behavioral sets. Some pharma organizations will fail to engage with this process, relying on the proven excellence and experience of their teams of qualified scientists. No doubt the advent of distributed, manufacturing elicited similar responses among the master craftsmen of the day. Other pharmas have already embraced the need for change and are engaging with many of the concepts described here. We shall watch their respective progress with interest.

References

1. J Walton. What Is the Investor View of Big Pharma Performance at Present? How Can the Industry Increase Value for Its Shareholders? Paper presented at Proceedings of R&D Leaders Forum, Geneva, Switzerland, Autumn 2002.
2. B Sherman, P Ross. The failure of industrialized research. *Acumen J Sci* 1:46–51, 2003.
3. IM Cockburn, RM Henderson. Publicity funded science and the productivity of the pharmacuetical industry: innovation policy and the economy. NBER Innovation policy and the economy. AB Jaffe, J Lerner, S Stern, Eds. Cambridge, MA: MIT Press, 1:1–34, 2001.
4. M Beggs. HTS: where next? *Drug Discov World* 1:25–27, 2000.
5. T Mander. Beyond uHTS: ridiculously HTS? *Drug Discov Today* 5:223–225, 2000.
6. S Udenfriend , LD Gerber, L Brink, S Spector. Scintillation proximity radioimmunoassay utilizing 125-I labeled ligands. *Proc Natl Acad Sci USA* 82:8672, 1985.
7. A Kolb, K Neumann, G Mathis. New developments in HTS technology. *Pharm Manuf Int* 31–39, 1996.
8. AJ Pope, UM Haupts, KL Moore. Homogeneous fluorescence readouts for miniaturized high throughput screening: theory and practice. *Drug Discov Today* 4:350–362, 1999.
9. RW Spencer. Diversity analysis in high throughput screening. *J Biomol Screen* 2:69–70, 1997.
10. K Rubenstein. Screening: trends in drug discovery. 2:34–38, 2001.
11. S Fox. *High Throughput Screening 2002: New Strategies and Technologies*. Moraga, CA: Hightech Business Decisions, 2002.
12. JA DiMasi, RW Hansen, HG Grabowski. The price of innovation: new estimates of drug development costs. *J Health Econ* 22:151–185, 2003.
13. S Fox, H Wang, L Sopchak, S Farr-Jones. HTS 2002: moving toward increased success rates. *J Biomol Screen* 7:313–316, 2002.
14. J Drews. The Impact of Cost Containment on Pharmaceutical Research and Development. Paper presented at Centre for Medicines Research Annual Lecture, Carshalton, Surrey, U.K., 1995.
15. SJ Arlington. *Pharma 2005: An Industrial Revolution in R&D*. Pricewaterhouse-Coopers Publications, 1999.
16. D Kell. Screensavers: trends in high throughput analysis. *Trends Biotechnol* 17:89–91, 1999.
17. RH Oakley, CC Hudson, RD Cruickshank, DM Meyers, RE Payne, SM Rhem, CR Loomis. The cellular distribution of fluorescently labeled arrestins provides a robust, sensitive, and universal assay for screening G protein-coupled receptors. *Assay Drug Dev Technol* 1:21–30, 2002.

18. WJ Harrison. The importance of automated sample management systems in realizing the potential of large compound libraries in drug discovery. *J Biomol Screen* 2:203–203, 1997.
19. S Holland. Breaking the Bottleneck in the Lead Discovery Process: The Development of the Automated Store (ALS) at Glaxo Wellcome. Paper presented at Proceedings of the 1998 Eurolab Automation Meeting, Oxford, pp. 31–31, 1998.
20. BA Kozikowski, TM Burt, DF Tirey, LE Williams, BR Kuzmak, DT Stanton, KL Morand, SL Nelson. The effect of freeze/thaw cycles on the stability of compounds in DMSO. *J Biomol Screen* 8:210–215, 2003.
21. BA Kozikowski, TM Burt, DF Tirey, LE Williams, BR Kuzmak, DT Stanton, KL Morand, SL Nelson. The effect of room temperature storage on the stability of compounds in DMSO. *J Biomol Screen* 8:205–209, 2003.
22. X Cheng, J Hochlowski, H Tang, D Hepp, C Beckner, S Kantor, R Schmitt. Studies on repository compound stability in DMSO under various conditions. *J Biomol Screen* 8:292–304, 2003.
23. M Snider. Screening of compound libraries … consommé or gumbo? *J Biomol Screen* 3:169–170, 1998.
24. TDY Chung. Screen compounds singly: why muck it up? *J Biomol Screen* 3:171–173, 1998.
25. JR Archer, L Wood. Production Tissue Culture by Robots. Paper presented at Proceedings of 11th Meeting of European Society for Animal Cell Technology, Wurzburg, Germany, 1993.
26. JR Archer, CT Matthews. Aseptic Manufacture of Live Viral Vaccines by Robotics. Paper presented at International Society of Pharmaceutical Engineers Proceedings, Basel, Switzerland, 1992.
27. AC Tee. The Use of Robotics for Large-Scale Production of Mammalian Cells or Their Products for HTS. Paper presented at Proceedings of ScreenTech 2000, Monterey, CA, 2000.
28. AC Tee. Automation of Large-Scale Adherent Mammalian Cell Culture. Paper presented at Proceedings of 7th Annual ESACT-UK Meeting, Birmingham, U.K., 1998.
29. M Montouté, M Fischer, S Lin. Application of Cell Culture Automation for Cell-Based HTS Using SelecT. Paper presented at Proceedings of European Society for Animal Cell Technology, Granada, Spain, 2003.
30. A Cacace. Industrialization of Cell Culture for Discovery. Paper presented at Society for Biomolecular Screening, Portland, OR, 2003.
31. M Beggs, JS Major. *Flexible Use of People and Machines in High Throughput Screening: The Discovery of Bioactive Substances*, J Devlin, Ed. New York: Marcel Dekker, 1997, pp. 471–481.
32. JS Major. A high performance automation platform. *Curr Drug Discov* 31–35, July 2002.
33. JA Mostacero. Asset: Industrial Automation for uHTS. Paper presented at Proceedings of Current Challenges in Biomolecular Screening, Madrid, November 13–14, 2003.
34. T Koppal. Robotics transform screening. *Drug Discov Dev* 30–34, 2003.
35. TC Peakman, S Franks, C White, M Beggs. Harnessing the power of discovery in large pharmaceutical organizations. Closing the productivity gap. *Drug Discov Today* 8:203–211, 2003.

36. JS Handen. High throughput screening: challenges for the future. *Drug Discov World* 47–50, Summer 2002.
37. PB Taylor, FP Stewart, DJ Dunnington, ST Quinn, CK Schultz, KS Vaidya, E Kurali, TR Lane, WC Xiong, TP Sherrill, JS Snider, ND Terpstra, RP Hertzberg. Automated assay optimization with integrated statistics and smart robotics. *J Biomol Screen* 5:213–225, 2000.
38. PD Haaland. *Experimental Design in Biotechnology*. New York: Marcel Dekker, 1989.
39. J-H Zhang, TDY Chung, KR Oldenberg. A simple statistical parameter for use in evaluation and validation of high-throughput screening assays. *J Biomol Screen* 4:67–73, 1999.
40. SW Armstrong. A review of high-throughput screening approaches for drug discovery. *Am Biotechnol Lab* 25–26, April 1999.
41. Evotec. EVOTEC OAI's uHTS System EVOscreen® Accepted by Pfizer US Evotec OAI, press release. Hamburg, November 29, 2000. Available at www.evotecoai.com/pls/portal30/docs/folder/evo_ca_gna/evo_fd_site_admin/evo_fd_pdf_lib/evo_fd_press_pdf/pr_evoscreeniib_291100_e.pdf.
42. Evotec. EVOTEC Ships Nanoliter Dispenser Unit DINA to SmithKline Beecham, press release. Hamburg, December 3, 1999. Available at www.evotecoai.com/pls/portal30/docs/folder/evo_ca_gna/evo_fd_site_admin/evo_fd_pdf_lib/evo_fd_press_pdf/pr_smithkline_dina_031299_e.pdf.
43. SA Sundberg. High-throughput and ultra-high-throughput screening: solution- and cell-based approaches. *Curr Opin Biotechnol* 11:47–53, 2000.
44. J Boguslavsky. Creating knowledge from HTS data. *Drug Discov Dev* 4:34–38, 2001.
45. JR Archer. Faculty or factory? Why industrialized drug discovery is inevitable. *J Biomol Screen* 4:235–237, 1999.
46. RR Ruffolo. R&D Productivity in the Pharmaceutical Industry: Issues and Challenges. Paper presented at Proceedings of IBC Drug Discovery Technologies Meeting, Boston, August 2003.
47. L Evans. The lab of the future: a drug discovery factory? *Curr Drug Discov* 23–26, 2003.
48. M Kaufman. Decline in New Drugs Raises Concerns: FDA Approvals Are Lowest in a Decade. *Washington Post*, November 18, 2002, p. A01.
49. CPFR. The Next Wave of Supply Chain Advantage CPFR Survey Findings and Analyst's Report, 2000. Available at www.cpfr.org.
50. D Garretson. Building Tier 0 Automotive Collaboration. Technology Strategy Forecast, Forrester Research, 2002. Available at www.forrester.com.
51. Genome Web. Despite R&D Spending Growth at Pharma, Tool Shops Will Be Left in the Cold. *GenomeWeb* (New York), February 27, 2003.

chapter 6

Parallel Lead Optimization

Steven L. Gallion, Alan Beresford, and Philippe Bey

Contents

6.1 Introduction

Drug discovery and development is notoriously difficult and expensive. Today, improving the productivity of R&D is unanimously recognized as the major challenge that the drug industry faces to maintain its historical growth rate. A number of recent assessments have described the inefficiencies within the industry and outlined many of the causes (1–4). It is now widely recognized that many of the risks carried late into development could, and should, have been dealt with in the earlier discovery phase. Starting in the mid-1990s, the motto in R&D became "fail fast" to avoid the costly late-stage failures in development by eliminating the losers quickly. The industry put in place high-throughput assays for ADMET (absorption, distribution, metabolism, excretion, and toxicity) to filter out the compounds with clear ADMET liabilities and developed predictive methodologies to select compounds with the least risk from virtual libraries before actual synthesis. The historic process of maximizing potency and selectivity prior to overcoming ADMET and other development liabilities is now superseded with a more balanced approach in which these criteria are examined much earlier (5,6). The expected outcome is a reduction of the cycle time and overall cost of bringing the drug candidates to the market.

Creating compounds for successful preclinical evaluation therefore involves concurrent investigation of multiple complex pathways. The navigation of these pathways is facilitated by the establishment of a therapeutic profile for the desired compounds, i.e., the specific characteristics needed for the therapeutic indication, a means of measuring progress toward this goal for both individual compounds and the parent chemical series, and a diverse set of appropriate leads (7,8). The selection of the preclinical candidate is the result of exploring each promising chemical series through a hierarchical cascade of tests, defined by the required therapeutic profile consisting of multiple, sometimes interdependent, parameters (Figure 6.1). The current paradigm in the industry is that a significant improvement in both success rate and efficiency will result when multiple lead series are investigated in parallel with simultaneous optimization of all relevant parameters. This review will discuss the current strategies and challenges in accomplishing this task.

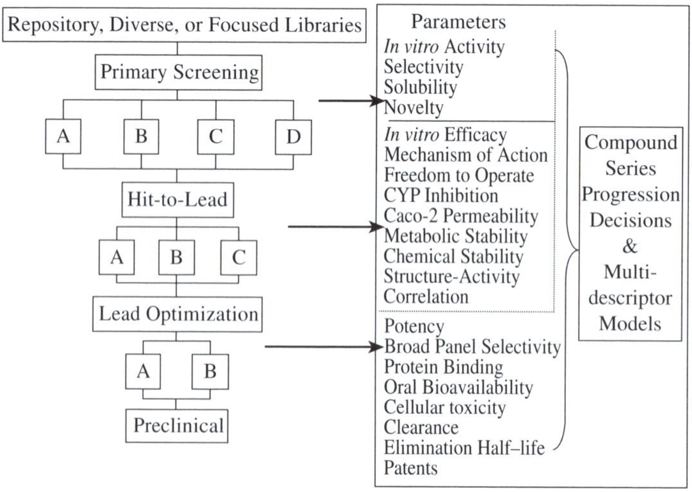

Figure 6.1 Lead discovery and optimization require many different components working in concert to achieve effective clinical results. The flow of multiple compound series through various phases is depicted on the left side of the diagram. At each stage decisions must be made based on all available data. Models created to explain and extrapolate each observable in turn rely on a variety of potential descriptors.

6.2 *Parallel Optimization Systems*

6.2.1 *Multiple Chemical Series*

The challenge of the discovery process is reflected by the attrition rates through the various phases. To maximize the chance of success for preclinical candidates, it is necessary to start with several different chemical categories of those hits and rapidly assess the characteristics of hits in terms of total drugability profile to select the lead series. Although there is some variation in terminology within the industry, a lead series is usually defined as one that demonstrates the following: dose-dependent activity *in vitro*, some degree of selectivity, reasonable structure-activity variation, novelty, and a validated mechanism of action (8).

Compound attrition rates differ greatly among programs and therapeutic areas, and estimates of attrition in discovery and preclinical development are difficult to obtain because no reliable sources exist. Success rates for new chemical entities completing the preclinical development stage are anywhere between 0.02 and 1% (9–12). In addition, it is estimated that no more than 4% of the candidates selected for good laboratory practice (GLP) toxicology studies reach the market (1,5,9–12).

Implicit in the fail-fast strategy is the assumption that backup compounds of better quality are immediately available to replace the losers. At a recent conference, one of the top five pharmaceutical companies reported that the attrition of their backup or second-generation compounds has been, in fact, higher than that of their first-generation compounds (3). Moreover, in the vast majority of the cases the backup or second-generation compounds fail for reasons that could not be anticipated from the knowledge derived from the first-generation compounds (Figure 6.2). This report might explain

New Candidates

Exposure	Safety/Tox	Efficacy	Other*
15%	46%	28%	10%

Backup Compounds

18%	55%	27%	0%

Applied Knowledge: Recurrence

No	Yes	Maybe
54%	8%	38%

* Other=Cost of goods, patent, "Portfolio Decision."

Figure 6.2 The need for diverse chemical series and the high degree of unpredictability during the preclinical and clinical development is supported by the attrition measured at one major pharmaceutical organization. Failure rates of the backups were similar to those of the first-generation compounds, and the reasons for failure were mostly unrelated to the liabilities identified with the first-generation compounds. (From F Douglas, Integrated Strategies to Increase R&D Productivity, paper presented at Drug Discovery and Technology Leaders Summit, Montreux, Switzerland, June 2003.)

why the R&D productivity, as measured by the number of new chemical entities approved every year, has continued to dwindle and points to a critical need to increase the number of lead series per target. It is therefore essential to retain multiple series during optimization. These backup compounds are required to have a good overall profile and sufficiently different chemical structures to have a reasonable chance of avoiding the same problems as the initial candidate. It is noteworthy that different chemical classes eventually emerge from different companies for the same target.

6.2.2 ADMET Considerations

6.2.2.1 Balanced Properties

A good medicine is a molecule that achieves the correct balance across a broad range of properties, including potency and selectivity for the therapeutic target, a good ADME profile, and minimal side effects at therapeutically effective doses. As has been acknowledged, achieving such a balance represents a lengthy and costly challenge for the pharmaceutical industry.

To expedite the process and reduce costs, the trend in the pharmaceutical industry has been to develop and implement high-throughput *in vitro* assays and *in silico* methods for ADMET predictions. These follow the success of such approaches for binding and selectivity optimization strategies. As yet, there is little evidence of increased productivity in either preclinical or phase I development, and there is a sense that the industry is trying to fit 21st-century components into a 20th-century machine and forfeiting much of the potential improvement in efficiency as a result (1–3,5,9–12). It could be argued that there has been insufficient time for these changes to yield their expected returns on the overall productivity. The "soft drug design" approaches, in which ADMET considerations are explicitly accounted for during optimization, have seen many selective successes (13,14), but extending the concepts into routine, systematized application still remains a challenge.

Clearly there is not, and will not be, a single drug discovery process. More likely, there will be variations around common themes that take into account the most appropriate scheme that fits the research and business strategy of each company. Many companies consider the optimization process completed when a candidate is selected for GLP toxicology studies, while others wait until proof of clinical efficacy in humans is established so that the clinical experience with first-generation candidates can be used for the optimization of the backup candidates. Historically, one of the important common themes has been the serial-cyclical process in lead optimization, where only compounds with optimal characteristics for property 1 (e.g., potency) are progressed for testing against property 2 (e.g., selectivity), and only those optimized for properties 1 and 2 progress to test 3 (e.g., absorption), and so on (15). The problem is that failure to find compounds satisfying the requirements of the later tests means restarting the process with a different series of compounds or accepting less than optimal performance in those later tests. And, the higher up the ladder a compound climbs, the more costly and frustrating the slide back down becomes. Hence, programs that set out with the intention of having nanomolar potency against their target receptor, 100-fold selectivity against related receptors, and > 50% oral bioavailability in animals may successfully achieve their potency criteria but end up accepting 50-fold selectivity and 25% bioavailability in their lead molecule.

The alternative to this serial process is a parallel one in which as many of the properties required of a medicine as possible are selected and optimized *at the same time* (16). This approach has the benefit of producing more balanced compounds, less likely to fall from the higher rungs of the ladder, and reduces the risk of having to repeat the entire process from the beginning. Computer modeling of ADMET properties, as reviewed elsewhere in this volume, can be combined with other predictive tools, such as activity and selectivity models, to calculate parameters for virtual compounds as part of a multidimensional optimization process. The ability to balance all the properties in an unbiased manner for many different compounds and to provide guidance within this complex space is beyond the skills of the most dexterous medicinal chemists. Hence, *in silico* predictive models offer an attractive means to address the problem. However, today even simple computer- based scorings of lead candidates using multiple criteria are often ignored by the discovery team (17), and one of the greatest challenges in research will be facilitating a cultural acceptance of *in silico* optimization schemes.

The ultimate goal is for the best combination of properties to be designed into a relatively small number of compounds at the outset or, rather, "to design the winners early" instead of "to eliminate the losers fast." Currently, the process relies principally on discarding unsatisfactory compounds from very large numbers of potential candidates. However, care needs to be taken to ensure that this weeding out process is a parallel one, particularly if it is performed *in silico* on virtual compounds. The advantage of being able to calculate or predict all the properties required in one impressive computational routine is lost if the output is simply the result of some sequential filtering algorithm. Understanding the causality of liability trends is more beneficial than collecting the output of the automated scoring process.

6.2.2.2 Early ADMET

Despite the realization by drug discovery scientists that potency is not the be all and end all for a good medicine, and that a compound needs to reach its intended target and remain there long enough to be effective, poor or inappropriate pharmacokinetics remains a major cause of failure for many compounds late in the process (18). To minimize the risk for compound progression and maximize the benefits of screening out molecules with poor ADME properties, pharmaceutical companies have been steadily moving pharmacokinetic (PK) and ADME testing to earlier stages of discovery (19).

These days, most large companies have ADME/PK groups whose function spans the whole discovery and development process, and many of those are now involved at the earliest exploratory/lead identification stages and in the design of drug-like compound libraries (20–22).

Moving toxicity testing earlier into the drug discovery process has so far proved to be a less tractable problem, and compound failure due to unwanted toxicity is still a major source of wasted resource. A major hurdle to moving toxicity assessment forward in drug discovery is that the causes and consequences of toxicity are various and variable. Furthermore, the toxic response observed may be the end result of a whole series of chemical and biochemical events that can only occur in an intact animal and may be dose and time dependent. Performing whole animal studies, involving multiple doses and requiring postmortem and histological analyses, is expensive with respect to time, resources, and quantities of compound required, i.e., not viable for a large number of compounds at an early stage of investigation. However, Safety Pharmacology guidelines, while recognizing the value of *in vitro* studies for studying the mechanisms underlying some toxic events, make it clear that compounds must be evaluated in intact, and preferably unanesthetized, animals (23,24).

For the purposes of this chapter, early ADMET is taken to mean any estimation or determination of ADME/PK properties and toxicity assessments carried out prior to, and in aid of, candidate selection, i.e., progression of a compound into full GLP-controlled toxicity studies. The definition may vary from company to company, but the tools available to both the ADME/PK scientist and toxicologist now cover the full spectrum from *in silico* through *in vitro* to *in vivo*. The choice of tool will depend on the number of compounds and resources available as well as the scientist's confidence in the ability of the tools to differentiate good from bad compounds.

It is reasonable to assume that confidence in this ability of the tools decreases as their apparent relevance to the clinical end-point decreases, i.e., confidence of results *in vivo* > *in vitro* > *in silico*. However, the available throughput these methodologies offer increases in the reverse direction. Hence, during lead optimization there is a natural shift in emphasis (from *in silico* to *in vitro* to *in vivo*) in going from the lead identification end to the candidate selection end of the process. An important feature in this shift is that the skill set of the user changes: as ADMET properties are considered ever earlier in the process, there tends to be increasing separation from those with specialized training. Indeed, the *in silico* methods are currently best suited to lead generation, where they allow ADMET properties to be taken into consideration even at the stage of virtual library design. Currently this is the domain of medicinal and computational chemists, and cross-disciplinary communication is vital in preventing misinterpretation of data. What constitutes an unacceptable ADMET profile for one therapeutic target may be perfectly acceptable in another area, or a profile acceptable early in a program may become less so as new data are generated and the program advances. Hence, if screens are used to select out compounds with unacceptable profiles, a clear understanding of how *and why* criteria change from project to project, or with time, is crucial.

6.2.2.3 ADMET Screening

When it comes to parallel optimization of ADME properties, *in vivo* PK studies represent the ideal way of comparing compounds, and various attempts have been made over the years to improve the throughput of *in vivo* animal studies and decrease the amount of compound required. These include a variety of techniques for obtaining serial blood samples from small animal species (25,26) and cassette dosing (27,28), where several compounds are administered simultaneously to the animals. The latter has been made possible by the development of mass spectrometry (MS) as a routine tool for bioanalysis (29), such that PK profiles can be monitored over reasonable periods using only microliter plasma samples with clinically relevant dose levels (28,30,31). Similarly, metabolites can often be identified in biological matrices from such studies by MS, although quantification is still largely dependent on the availability of radiolabeled compounds. Full characterization by nuclear magnetic resonance (NMR), though greatly improved in recent years, is still not possible in the same concentration range (32,33).

The main problems with *in vivo* studies in drug discovery are that (1) they cannot be performed in the end-user species, man, for experimental compounds, and (2) if the compounds tested do not have the desired PK profile, it can be extremely difficult to dissect out the cause(s) of the problem. For example, if compounds show consistently poor oral bioavailability, this may be due to low stability or solubility in the GI tract, poor absorption across the gut wall, extensive metabolism of absorbed compound in the gut wall or liver (by a variety of enzyme systems), extensive excretion into the bile, and so on. Consequently, it helps to have systems in place that can screen for each of these potential problems in parallel and provide insight into the structure-activity relationship (SAR) governing each of these properties. Ideally, each of these screens would be a model system for man. Of course, while breaking down the complex whole into simpler, more manageable sections can help our understanding of the rules underlying various ADME properties, this brings with it the problem of understanding how to integrate the sections. As previously mentioned, the simpler and more mechanistic the model, the less relevant the model seems to the *in vivo* situation, and often the less likely it is to be used as a decision-making tool for compound progression.

Surgical procedures have been developed to try and isolate a number of the potential variables from whole animal PK studies, including cannulation of specific organs, ducts, and blood vessels, such as the bile duct (34) or hepatic portal vein (35–37), and the insertion of gastric fistulae (38,39). Alternatively, *in situ* isolated organ perfusion methods have been developed that include the heart (40), lungs (41,42), kidneys (43), liver (44), intestines (45,46), and brain (47).

In rare cases, mutant strains of animals may have characteristics that, by comparison with wild-type animals, can help to isolate particular ADME mechanisms. For example, the inbred bilirubin-UDP-glucuronosyl-transferase- deficient Gunn rat can be used to demonstrate the involvement of this enzyme in conjugation of certain substrates (48). In recent years, transgenic and genetic knockout mice have been developed for a number of metabolizing enzymes (49,50) and transport proteins such as p-glycoprotein (Pgp), but again, the relevance of these to humans remains debatable (51).

Compared to many of the above, the relative simplicity and reproducibility of *in vitro* methods make them better suited to routine screening. In addition, many of these methods are amenable to automation, and hence higher throughput, and can often be based on human-derived material (52). Consequently, these tend to be the first tools of choice in modern ADME/PK laboratories.

6.2.2.3.1 Absorption Screens. The determination of solubility and stability of novel compounds at different pH values and in simulated gastrointestinal fluid can help to identify potential problems with oral bioavailability (53). The Caco-2 cell system has become a widely accepted model for intestinal absorption (54). Although similar data may be obtained using other cell lines that are easier to work with and maintain, such as the Madin–Darby canine kidney (MDCK) cell line (55,56), much of the popularity of Caco-2 is due to its origin as a human intestinal cell line, hence appearing more relevant to the intended end user. Although such cell lines are primarily used to estimate passive diffusion of compounds across the gut, they do express transport proteins, notably Pgp. Consequently, comparison of bidirectional passage across the cells, studies of temperature dependency, or the use of transport inhibitors can provide valuable information on the contribution of active transport to uptake or efflux of compounds in the gut (56). Indeed, there is increasing focus on the use of these systems, particularly MDCK cells, to study active transport by engineering overexpression of transporter proteins (57). Alternatively, simplified systems have been developed to isolate the passive component of absorption using artificial membranes such as the filter immobilized artificial membrane (IAM) technology and parallel artificial membrane permeation assays (PAMPAs) (46,58,59).

6.2.2.3.2 Distribution Screens. The measurement of tissue distribution is still largely restricted to *in vivo* studies, using radiolabeled material or the application of extensive extraction and cleanup procedures for tissue homogenates to ensure quantitative recovery of compounds for analysis. The drawback of such measurements is that they frequently use whole tissue homogenates, and whether the compound resides in specific, relevant fractions of those tissues remains unknown. The possibility of poor penetration of compounds into target cells is a common concern, particularly when a significant decrease in potency is observed in going from primary screening against isolated enzymes/receptors to secondary cell-based screens. Again, while specific analytic methods can be developed to measure compounds incubated in cell systems, there often remains a question as to exactly where the compound is localized within the cell or if it is bound to the cell wall. Of increasing importance in ADME studies is the influence of transporter systems on drug distribution. Uptake and efflux of molecules at the blood–brain barrier (BBB) is of particular interest because of the potential effects in preventing effective penetration of drugs intended for central nervous system (CNS) targets or side effects for compounds not intended to penetrate the CNS. As with absorption of drugs from the gut, despite the existence of a variety of active transport systems, Pgp remains the most widely studied, and a cell coculture-based BBB model has been developed that expresses this protein (60).

Tissue and plasma protein binding are the major factors influencing the distribution of compounds, and hence both their efficacy and ADME characteristics. Plasma protein binding is by far the most widely studied and characterized, not least because of its accessibility. It can be readily determined *in vitro* at relatively high throughputs using whole plasma from humans or animals, or specific isolated proteins (61). Percentage binding is most commonly measured, as it is considered that only the fraction of drug unbound is available for activity, as well as for distribution and clearance. However, the binding affinity and the equilibrium rates for binding to both plasma and cellular proteins are of major importance. Determination of affinity and rate constants is more time consuming, although measurements can be done in some 96-well format assays or using immobilized protein high performance liquid chromatography (HPLC), columns to provide relatively high throughput. Affinity, rather than percentage bound, may be particularly important if it potentially restricts access to metabolizing enzymes, or sites of activity such as the CNS. Affinity and rate of equilibrium may be important if the site of action is directly accessible to the plasma, as with thrombin or factor Xa inhibitors. So, while percentage plasma protein binding is a routinely measured and compared value for experimental compounds, understanding the significance of the values for ADME, or efficacy, is by no means straightforward. Indeed considerable misunderstanding still exists regarding the potential for drug–drug interactions resulting from the displacement of one drug bound to plasma protein by a coadministered compound. While this displacement may be demonstrated *in vitro*, the dynamics of this process, coupled with redistribution and potential clearance changes *in vivo*, make the clinical relevance of these effects difficult to assess. In any case, potential drug–drug interaction factors resulting from protein displacement are less important than those involving alterations in absorption, metabolism, or renal excretion (62).

6.2.2.3.3 Metabolism Screens. As well as understanding the metabolism of experimental compounds, it is also important to try to gauge their potential effect on the metabolism of other, coadministered compounds and endogenous molecules. Hence, metabolic turnover, enzyme induction, and enzyme inhibition are all important factors to assess.

There is now a plethora of *in vitro* systems available for studying metabolism and the interaction of compounds with metabolizing systems. As the major organ of clearance, the liver is of primary interest (reviewed in 63), the P450 cytochromes and the P450 cytochrome enzymes in particular (reviewed in 64). As with Caco-2 cells, the use of liver slices and hepatocyte suspensions is particularly popular because human tissue is available from a number of surgical procedures and there is a certain comfort factor in results obtained from what is considered to be a relatively complete metabolism system (65), i.e., capable of carrying out both phase I (oxidative) and phase II (conjugative) metabolism. Cultured hepatocytes and human liver slices can also be used to study the inhibition and induction of metabolic enzymes and transport proteins (66,67). Unfortunately, although hepatocyte-based screens are considered highly relevant for modeling metabolism in man, these are technically challenging to use routinely. Hence, simpler, more reproducible subcellular fractions such as S9 and microsomes are more widely used for medium- to high-throughput screening for metabolic stability (52). These fractions can also be prepared from other tissues, such as gut and lung, if there are particular concerns over metabolism in these tissues, although enzyme levels are usually too low to make these useful on a routine screening basis. Again, there are systems intended to further dissect out the underlying metabolic routes and mechanisms, with many human metabolic enzymes now commercially available, along with a variety of probe substrates and inhibitors of varying specificity and selectivity for each enzyme. The relative simplicity and reproducibility of these systems make them ideal for high-throughput screening for both metabolic turnover and inhibition studies (19,68), to the extent that data processing and interpretation can become a bottleneck.

Induction of metabolic enzymes is potentially a "kill" criterion in drug development, although there are many compounds on the market that are inducers. Inducers of the CYP1A family have received particular attention over the years because of their association with activation of carcinogens (69), but many natural products and dietary constituents have this effect, and compounds such as omperazole, for which CYP1A induction is reported (70), have been highly successful drugs. However, enzyme induction can result in increased clearance, and hence reduced efficacy, of coadministered drugs, or of the compound itself if a substrate for the enzyme induced. If extensive, induction may result in a significant increase in liver weight and may have serious consequences for the metabolism of endogeneous molecules. Historically, testing for induction has required *in vivo* animal studies (71), and the relevance of this to humans, particularly for P450s, where the isoforms present show considerable species differences, has been difficult to assess. However, as noted above for hepatocytes, there has been considerable success in demonstrating induction *in vitro*. Modern techniques such as real-time reverse transcription polymerase chain reaction (PCR) have been used to examine P450 induction in liver slices (72), and the identification of nuclear receptors such as PXR and CAR as key regulators of P450 and possibly drug transporter gene expression (73) has led to the development of high-throughput reporter gene assay screens measuring the interaction of new compounds with these receptors (74). However, the direct relevance of this to *in vivo* induction has yet to be clearly established (75).

6.2.2.3.4 Excretion Screens. The major routes of excretion of compounds, and their metabolites, from the body are by the urine and the feces (principally via the bile). These processes require the intact kidney and liver, and consequently *in vivo* studies. Passive elimination of unchanged drug by either of these routes is relatively slow and will be dependent on plasma protein binding. However, active secretion can have a huge impact, particularly for bile, where a significant proportion of an orally absorbed dose may be cleared in its first pass through the liver. As for absorption and distribution, increased understanding of the role of transporters in active excretion of drugs will help in future compound design, but currently the main hope for significant improvement lies in the development of *in silico* models, which may be able to predict the likelihood of excretion based on physicochemical properties and existing experimental data.

6.2.2.3.5 Toxicity Screens. As previously noted, toxicity is frequently a multifactorial event, and the possible toxic responses are almost limitless: from skin irritation and lachrymation, through neurotoxicity, to cancer and teratogenicity. Similarly, the mechanisms underlying the observed toxicity may be simple or extremely complex: from the direct effect of some corrosives to complex chains of biochemical events, including metabolic activation of the potential toxin within the body. Consequently, many early toxicity tests involve whole animal *in vivo* studies, but are designed to look for specific effects that have already been reported for related compounds or compounds in a particular therapeutic category, for example, monitoring for changes in the QT interval associated with blockage of the cardiac potassium ion current through the human ether-a-go-go-related gene (hERG) channel (76–78). Other whole animal studies include those for P450 induction previously mentioned, and observations of generalized posture and behavioral changes, as in the Irwin's test or the mouse defense test battery (79). However, many of the systems and techniques applied in ADME have also been used to develop *in vitro* toxicity test systems. Hence, for example, hepatotoxicity has been studied in isolated, perfused organ models, liver slices, and hepatic cell systems (80).

Perhaps the most common and long established *in vitro* toxicity tests have been those developed for genotoxicity (reviewed in 81–83), including the Ames test (84–86) and, more recently, the COMET assay, the micronucleus assay, and the hypoxanthine–guanine phosphoribosyltransferase (HPRT) mutant assay (87–90). The continued improvement in cell culture technology has increased the number of tests available, with a variety of different end-point measurements to help answer questions about the mechanisms of toxicity (91). High-throughput *in vitro* cytotoxicity assays are now routinely available (92,93), and where the upregulation of specific genes has been historically associated with particular toxicities, marker assays have been developed using reporter gene constructs (94,95). DNA chip technology (96–99), a natural progression of this technique, provides a means of early toxicity assessment and potentially offers new insights into mechanisms of both toxicity and disease progression (100). The use of these technologies has spawned the word *toxicogenomics*, which in its broadest definition also includes monitoring for changes in the products of gene expression, including messenger RNAs (101). The expectation is that characteristic changes, not only in the pattern of expressed genes, but also in proteins (proteomics) (102) or cellular metabolic products (metabonomics) (103,104), can be identified that are indicative of particular toxic events. The hope is that these changes will become apparent in samples taken from animals after relatively short term exposure to the compound under investigation, or that the changes can be identified using *in vitro* systems. In the short to medium term, however, considerable work will be required to validate these systems by comparing their results to those obtained by conventional animal tests with full postmortem and histological analysis.

The above summary of available ADMET screens, which is by no means exhaustive, indicates that a vast amount of money and resources can be expended on early ADMET testing. This is one of the reasons why testing has largely remained a sequential rather than parallel affair. So, it is hardly surprising that drug discovery scientists have put increasing effort into the development of predictive computational ADMET models to reduce the amount of practical experimental testing required (105).

6.2.3 Computational Methods

The past decade has seen extensive development and application of computational technologies within the entire spectrum of the drug discovery process. Part of the driving force has been the increasingly large numbers and varieties of chemicals produced concomitantly with the expansion in assay and target technology. The volume of data generated has resulted in significant development of both data management systems and data analysis procedures. Computational chemistry tools have evolved to assist in many classical aspects of this analysis, and there are a number of excellent reviews covering these methods in detail (106–108). Likewise, models for predictive ADMET parameters have been reviewed elsewhere in this volume.

To parallelize the lead optimization at the molecular level, one must address the challenges of (1) how to translate the output of these ADMET predictive models in a manner conducive to prospective design (i.e., meaningful to chemists) and (2) how to best combine ADMET values (either experimental or calculated) with values for target affinity and selectivity. If these challenges can be solved, molecular design will become less a matter of serialized, one-variable-at-a-time optimization and become truly parallel. We begin by discussing typical metrics used to construct models with respect to providing information for synthetic and testing decisions. The second part discusses the merging of multiple parameters to make concerted alterations within or between series.

6.2.3.1 Metrics in Parallel Design

Acquiring novel, multiple chemical series is fundamental to the success of any therapeutic program, and accordingly, a great deal of emphasis has been given to assessing the diversity of compound collections. A number of methods have evolved to assist in file enrichment and are routinely used in the industry (109–113). While achieving diversity within a screening collection is desirable, optimization generally entails using the first synthetic cycles to make focused sets of compounds around confirmed hits with the expectation of producing a reasonable sampling of multiple physiochemical, ADMET, and SAR properties for each individual series. The use of experimental design methods ensures efficient sampling of the parameters and appropriate selection of the best metrics. In pursuing multiple chemical series in parallel, it is important to make certain that each of the series has sufficient quality and quantity of data in order to make informed decisions in a consistent manner (114,115). Other important factors, such as synthetic feasibility (particularly in parallel synthesis where reactivity differentials may be important), time, and cost, must also be considered (116).

To balance the physiochemical diversity with the appropriate ADMET considerations, many metrics have been used to characterize the drug-like or lead-like nature of libraries (117–129). These descriptors ranged from simple, classical metrics such as logP, molecular weight (MW), and number of H bond donors and acceptors to the more esoteric, e.g., autocorrelation vectors, atom layers, and the Altenburg polynomial. Although many of these descriptors provide the basis for models having good correlation, translating the output into effective prospective design can be daunting.

Many different classical descriptors and methods are available for parallel lead optimization. The Topliss decision tree is one example of simple, but powerful, fragment and position classifiers to guide SAR development, which has evolved to accommodate ADMET considerations (130). Likewise, molecular weight, log P, log D, and electronic or steric constants to guide synthesis are intuitive, easily implementable concepts (14,131,132). Small sets of descriptors, such as in the noted rule-of-five model, are routinely used to select for increased oral bioavailability (133). The rule of five specifies that no more than two boundary conditions should be violated for absorption by oral administration: molecular weight < 500, hydrogen bond donors < 5, hydrogen bond acceptors < 10, logP < 5. Simple physical properties are still very effective ways of measuring compound trends in development (134).

Models for specific ADMET end points can also be quite complex and, in many cases, utilize three-dimensional fields, pharmacophores, or more abstract descriptors, such as those derived from wave functions or graph theory. Examples include the work with intestinal absorption, blood–brain barrier penetration, CYP450s, and hERG (135–139).

A significant advantage of three-dimensional field and pharmacophore models is the relative ease to communicate SAR information to the medicinal chemist. The consequences of placement of features such as H bond acceptors and donors, aromatic rings, and excluded volumes are intuitive and conceptually facile. This strategy has also been effectively utilized to construct target-biased libraries (140).

The ability to combine multiple models to execute concerted changes in independent parameters remains difficult. Incorporating selectivity information in cases where crystallographic information is available has proven successful in designing kinase inhibitors (141). Extrapolating the results to cases where atomic level resolution is poor or nonexistent is quite difficult, as errors in the models are large in comparison to the energy differences needed to achieve good selectivity. Likewise, a number of toxicology models involve rule-based inference mechanisms that may generate results that are sometimes not easily interpreted in terms of substructure, although the use of recursive partitioning, toxophores, and **phylogenetic-like trees** in model derivation provides effective feedback (142–145).

The rules derived from simple descriptors are, for the most part, conceptually concrete and can be used not only in computational filtering virtual libraries, but also in the optimization of individual compounds. Every medicinal chemist is able to directly translate these rules into specific modification of substructures. If the essential SAR features are known, drug-likeness may be enhanced without unduly affecting activity, or even combined activity and selectivity parameters. However, if the ADMET property is not a simple function of a few tangible descriptors, computational models must be used. The challenge is to provide the information in a manner that is easily translatable into chemical modification. This becomes increasingly important if there is more than one problem to overcome in a lead compound. For example, if eliminating a hydrogen bond donor is expected to aid passive absorption, how will it affect metabolism by 3A4? The problem quickly becomes intractable as more dimensions are added. Even if highly predictive models are used simultaneously to score a particular idea, without interpretation and guidance, the process of design degenerates into a trial-and-error sampling. Rank ordering of compounds on the basis of scoring virtual libraries is a form of automated trial and error and is unfortunately the state of the art. Better trend analysis or pattern recognition methods are needed to facilitate conceptual understanding and development of innovative molecular designs.

6.2.3.2 *Multidimensional Optimization*

The initial efforts at establishing SAR within a series are to rationalize the effects of structural variation on a variety of experimental observables, as well as the metrics most relevant to those observables. Visualization tools play a critical role in the assessment of the comparative values for multiple properties in several different series (146). Simple pair-wise plots using spreadsheet programs can be used to accomplish much of the visualization, whereas use of statistical packages provides further analysis of intervariable correlations.

These analyses are quite effective in identifying trends related to global molecular properties. For example, plotting polar surface area (PSA) and MW for multiple sets of compounds and comparing them to a set of standards is effective in estimating the probability of absorption (14). However, without corresponding information on structure, activity, and other pharmacokinetic parameters, it will rarely be obvious what specific modification is to be made to result in a molecule or set of molecules to obtain an overall better profile.

Diagrams called flower plots have been developed to map several properties by R-groups (115). Trends can certainly be noted when these flower plots are used in conjunction with substructures and activity data, though it requires significant effort to interpret the density of information. A number of commercial programs allow several parameters to be displayed simultaneously with structural information (147–150,162). The capability to organize and display results in a meaningful manner by R-group variation is especially useful to provide perspective and analyze trends. While routinely performed within a single lead series, this analysis can also have distinct value across series if the structure-activity or structure-property trends are convergent.

Some data mining techniques such as phylogenetic-like classification and statistical analyses of R-group populations are convenient tools for elucidation of initial SAR (145,151). It would be interesting to explore use of these approaches where several parameters are involved. Neural nets have been effectively used to discriminate between drugs and nondrugs, as well as other bimodal classes; however, the approach lacks interpretative ability. If the virtual libraries for all series are exhaustive, then this remains a quite useful technique for focusing on higher-value compounds.

Oprea (126) summarized three possible strategies to optimize ADMET and binding parameters. The first is the classical approach of postponing ADMET until a set of potent, selective compounds has been identified. Most experienced medicinal chemists, however, do involve pharmacokinetic considerations to some extent during the early lead optimization process by adhering to established guidelines for creating effective compounds. Although not ignored, ADMET does tend to be deprioritized at the expense of novelty, potency, and selectivity, particularly during early lead discovery and optimization.

The second strategy is to apply a set of computational ADMET filters prior to synthesis. This is certainly an effective strategy for providing enrichment over the preceding cycle, but many potentially potent and selective compounds with poor, but optimizable ADMET profiles are precluded from discovery. All models are imperfect, and depending on the phase of the project and extent to which the models have been validated for the series of interest, it is customary to also include some percentage of compounds that will enhance SAR information or further test the applicability of the models, even though the compounds fall below the thresholds of the filters.

The third strategy is to perform simultaneous improvement in ADMET, potency, and selectivity. The method seeks to focus only on series and compounds with the lowest liabilities and highest potencies at any given point in the lead optimization process. Concurrent optimization is typically complicated by conflicting molecular determinants, for example, a hydrogen bond conferring selectivity against a related receptor but deterring from absorption characteristics. If the problem is noted early in the lead optimization process, there is a broader spectrum of available options to solve it. However, a conflicting concern is that the resolution of structure-property and structure-activity relationships may be insufficient to provide definitive concerted direction during the early stages of lead optimization.

There are a number of examples employing simultaneous optimization of multiple objectives with the search space composed of virtual libraries. One effective strategy has employed the design of focused lead optimization libraries with increased absorption and affinity profiles using rule-of-five and PSA descriptors together with synthetic constraints (128,151). It was noted that the use of easily interpretable metrics facilitated appropriate synthetic modifications. Among the synthetic constraints employed was the requirement that libraries be fully combinatorial for maximum efficiency.

A similar strategy has been employed by Bravi et al. (152) to optimize selection of a prefiltered reagent set. Optimizations of functions containing various diversity and ADME metrics have been performed (153–155). Each of these examples has focused on sequential rounds using a filtration or weighted sum fitness function. Shimada et al. (156) reviewed the process and discussed a typical compound profile chart used to prioritize compound progression.

A somewhat different approach was pursued to locate leads within a biased ion channel library comprised of several different chemical classes (157). Consensus scores of the predictive ADMET components for a series of 32 probe libraries were used in conjunction with SAR, novelty, selectivity, and experimental early ADMET information to prioritize series and to select compounds for synthesis from subsequent virtual libraries (158). The weights for each parameter were determined according to the relative importance of each parameter in the final objective therapeutic profile. Each weight was scaled based on the accuracy of the model, and the total score for each compound was defined as the product of the individual components. No attempt was made to optimize the ADMET profiles during the first two iterations. The first

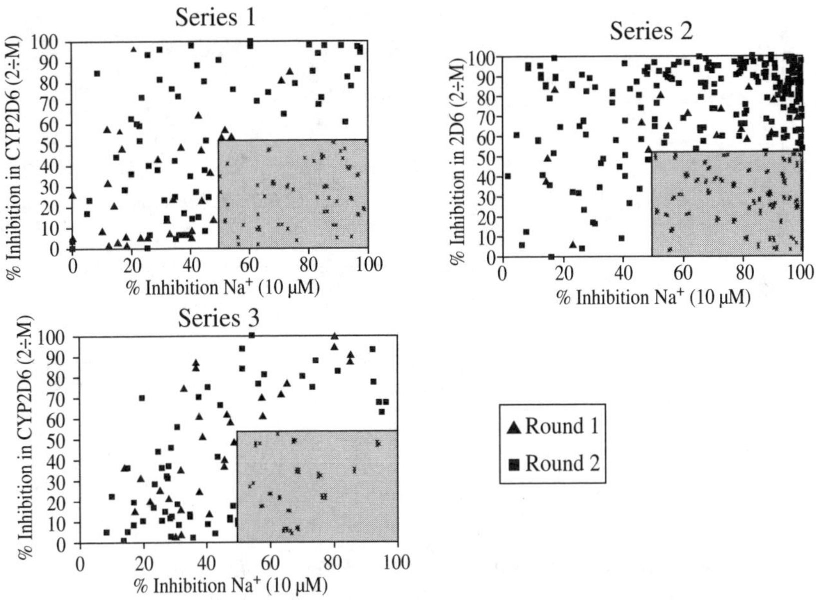

Figure 6.3 This shows one example of comparing multiple chemical series by plotting one risk (CYP450 2D6 affinity) as a function of affinity to the primary target (Na^+ channel). The region of lower-risk compounds is shaded. The results of two rounds of screening are shown. The first round consisted of a small probe set of compounds, while those in the second round were synthesized to help establish a structure-activity relationship. The first two series show the best balance of properties. (From SL Gallion, Biased Library Design and Parallel Optimization of Ion Channel Modulators, paper presented at International Business Conferences: Advances in Structure-Guided Drug Discovery, San Diego, CA, November 2002.)

iteration comprised a small diverse probe set. The second iteration consisted of a small set of compounds synthesized to map SAR. Inhibition of CYP2D6 was a significant risk for many compounds in most series. Figure 6.3 shows the results for the first two iterations of three independent series for the Na^+ channel, plotting binding affinity for the primary target against that of CYP2D6 at the primary screening concentrations.

The third iteration for series 2 involved four subseries, varying the central scaffold in an attempt to reduce CYP2D6 liability while preserving affinity to the Na^+ channel. A quantitative structure-activity relationship (QSAR) model was used to predict binding affinity, and the ArQule predictive ADME tools were used to predict ADME properties. Each virtual library was scored and a subset of compounds made. The results summarized in Figure 6.4 show that two of the subseries have reduced liabilities for CYP2D6 while retaining binding affinity to the receptor (159).

While the use of consolidated fitness functions for performing concerted changes in multiple parameters has proven useful, there are a number of issues,

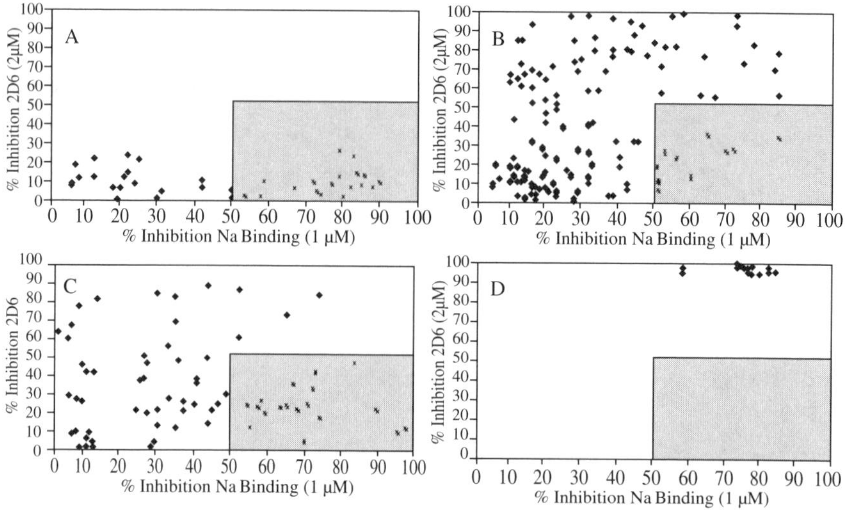

Figure 6.4 The results for a third iteration of optimization for one chemical series are shown grouped by structural subclass. In this case, subseries A and C contain the greatest density of compounds with the best set of balanced properties. (From SL Gallion, Biased Library Design and Parallel Optimization of Ion Channel Modulators, paper presented at International Business Conferences: Advances in Structure-Guided Drug Discovery, San Diego, CA, November 2002.)

including how to establish an *a priori* weighting system and, with additive functions, how to prevent a few high scores in certain parameters from unduly biasing the result. The use of Pareto scoring as the basis for optimization has been proposed to alleviate empirical bias and facilitate identification of multiple alternative solutions (160). A compound is Pareto optimal if an increase in one parameter does not lead to a decrease in another parameter. The process therefore provides a set of solutions for the experimentalist to select from, preserving compounds that strict filters would remove from further consideration. A combination of experimental design and Pareto optimization methods should greatly facilitate a more balanced exploration of compound space (161). However, to solve individual problems and provide directionality in multidimensional optimization beyond the scope of the initial virtual library, computational methods will be required to confer an additional degree of understanding as to the causality and interrelationships between structure and function.

6.3 Conclusion

There have been a large number of tools developed over the past decade in attempts to expedite the delivery of more effective compounds to the clinic. Although the success rate in clinical trials has not apparently increased to date, the increasing awareness and effort expended to produce novel, useful chemistries and to optimize all aspects of the required therapeutic profile should eventually be successful. The increasingly parallel nature of the discovery process is an important component of this effort. Multiple chemical series, decision criteria, models, metrics, and processes all contribute to locating and successfully optimizing new chemical entities. Accessing novel, effective chemistry in a timely manner will be another ongoing challenge for the industry. Continuing efforts in developing additional therapeutically relevant animal models, assays, and *in silico* models are needed, as are methods of extracting and presenting information to the project team in an efficient manner. The ability to extrapolate from *in silico* experiment to human trials remains a formidable challenge. As the science evolves, it is hoped that future lead optimization activities will be guided less by intuition and more by information.

Acknowledgments

We thank Drs. David Hartsough and Alan Hillyard for their comments on the manuscript and Ms. Judy Blaine for her aid in assembling and formatting the references.

References

1. T Kennedy. Managing the drug discovery/development interface. *Drug Discov Today* 2:436, 1997.
2. *Outlook 2002*. Boston: Tufts Center for the Study of Drug Development, 2002.
3. F Douglas. Integrated Strategies to Increase R&D Productivity. Paper presented at Drug Discovery and Technology Leaders Summit, Montreux, Switzerland, June 2003.
4. EL Allan. Balancing quantity and quality in drug discovery. *Drug Discov World* 4:71–75, 2002/2003.
5. JA DiMasi. New drug development in the United States from 1963 to 1999. *Clin Pharmacol Ther* 69:297–307, 2001.
6. RA Lipper. E pluribus product: how can we optimize selection of drug development candidates from many compounds at an early stage? *Mod Drug Discov* 2:55–60, 1999.
7. RA Goodnow, Jr. Current practices in generation of small molecule new leads. *J Cell Biochem Suppl* 37:13–21, 2001.
8. AD Baxter, PM Lockey. "Hit" to "lead" and "lead" to "candidate" optimization using multi-parametric principles. *Drug Discov World* 2:9–15, 2001.
9. GW Caldwell. Compound optimization in early and late phase drug discovery: acceptable pharmacokinetic properties utilizing combined physicochemical, *in vitro* and *in vivo* screens. *Curr Opin Drug Discov Dev* 3:30–41, 2000.
10. RA Edwards, K Zhang, L Firth. Benchmarking chemistry functions within pharmaceutical drug discovery and preclinical development. *Drug Discov World* 3:67–71, 2002.

11. JA DiMasi. Uncertainty in drug development: approval success rates for new drugs. In *Clinical Trials and Tribulations*, 2nd ed., revised and expanded, AE Cato, L Sutton, A Catto III, Eds. New York: Marcel Dekker, 2002, chap. 20, pp. 361–377.
12. DB Lakins. Nonclinical drug development: pharmacology, drug metabolism, and toxicology. *New Drug Approv* 100:17–54, 2000.
13. N Bodor, P Buchwald. Soft drug design: general principles and recent applications. *Med Res Rev* 20:58–101, 2000.
14. H van de Waterbeemd, DA Smith, BC Jones. Lipophilicity in PK design: methyl, ethyl, futile. *J Comput Aided Mol Des* 15:273–286, 2001.
15. AP Beresford, HE Selick, MH Tarbit. The emerging importance of predictive ADME simulation in drug discovery. *Drug Discov Today* 7:109–116, 2002.
16. I Kariv, RA Rourick, DB Kassel, TDY Chung. Improvement of "hit-to-lead" optimization by integration of *in vitro* HTS Experimental models for early determination of pharmacokinetic properties. *Comb Chem High Throughput Screen* 5:459–472, 2002.
17. SJF MacDonald, PW Smith. Lead optimization in 12 months? True confessions of a chemistry team. *Drug Discov Today* 6:947–953, 2001.
18. GW Caldwell, DM Ritchie, JA Masucci, W Hageman, Z Yan. The new preclinical paradigm: compound optimization in early and late phase drug discovery. *Curr Top Med Chem* 1:353–366, 2001.
19. PJ Eddershaw, AP Beresford, MK Bayliss. ADME/PK as part of a rational approach to drug discovery. *Drug Discov Today* 5:409–414, 2000.
20. AR Leach, MM Hann. The *in silico* world of virtual libraries. *Drug Discov Today* 5:326–336, 2000.
21. H Matter, KH Baringhaus, T Naumann, T Klabunde, B Pirard. Computational approaches toward the rational design of drug-like compound libraries. *Comb Chem High Throughput Screen* 4:453–457, 2001.
22. WJ Egan, WP Walters, MA Murcko. Guiding molecules toward drug-likeness. *Curr Opin Drug Discov Dev* 5:540–549, 2002.
23. LB Kinter, JP Valentin. Safety pharmacology and risk assessment. *Fundam Clin Pharmacol* 16:175–182, 2002.
24. DJ Snodin. An EU perspective on the use of *in vitro* methods in regulatory toxicology. *Toxicol Lett* 127:161–168, 2002.
25. BH Migdalof. Methods for obtaining drug time course data from individual small laboratory animals: serial microblood sampling and assay. *Drug Metab Rev* 5:295–310, 1976.
26. WM Williams. A chronically implantable arterial catheter for use in unrestrained small animals. *J Neurosci Methods* 12:195–203, 1985.
27. RE White, P Manitpisitkul. Pharmacokinetic theory of cassette dosing in drug discovery screening. *Drug Metab Dispos* 29:957–966, 2001.
28. WA Korfmacher, KA Cox, KJ Ng, J Veals, Y Hsieh, S Wainhaus, L Broske, D Prelusky, A Nomeir, RE White. Cassette-accelerated rapid rat screen: a systematic procedure for the dosing and liquid chromatography/atmospheric pressure ionization tandem mass spectrometric analysis of new chemical entities as part of new drug discovery. *Rapid Commun Mass Spectrosc* 15:335–240, 2001.
29. PM Timmerman, R de Vries, BA Ingelse. Tailoring bioanalysis for PK studies supporting drug discovery. *Curr Top Med Chem* 1:443–462, 2001.

30. BL Ackermann, MJ Berna, AT Murphy. Recent advances in use of LC/MS/MS for quantitative high-throughput bioanalytical support of drug discovery. *Curr Top Med Chem* 2:53–66, 2002.
31. T Ohkawa, Y Ishida, E Kanaoka, K Takahashi, H Okabe, T Matsumoto, S Nakamoto, J Tamada, M Koike, T Yoshikawa. A new generic column switching system for quantitation in cassette dosing using LC/MS/MS. *J Pharm Biomed Anal* 31:1089–1099, 2003.
32. SS Pochapsky, TC Pochapsky. Nuclear magnetic resonance as a tool in drug discovery, metabolism and disposition. *Curr Top Med Chem* 1:427–441, 2001.
33. AP Watt, RJ Mortishire-Smith, U Gerhard, SR Thomas. Metabolite identification in drug discovery. *Curr Opin Drug Discov Dev* 6:57–65, 2003.
34. LL Rolf Jr, KE Bartels, EC Nelson, KD Berlin. Chronic Bile duct cannulation in laboratory rats. *Lab Anim Sci* 41:486–492, 1991.
35. M Bickel, HF Brettel, S Jung, R Schleyerbach. A New model for demonstrating the hepatic first-pass effect. *Arzneimittelforschung* 27:1879–1881, 1977.
36. K Tabata, K Yamaoka, T Fukuyama, T Nakagawa. Evaluation of Intestinal absorption into the portal system in enterohepatic circulation by measuring the difference in portal-venous blood concentrations of Diclofenac. *Pharm Res* 12:880–883, 1995.
37. JH Strubbe, JE Bruggink, AB Steffens. Hepatic portal vein cannulation for infusion and blood sampling in freely moving rats. *Physiol Behav* 65:885–887, 1999.
38. RG Sample, GV Rossi, EW Packman. Thiry-Vella dog as a biologic model for evaluation of drug absorption from the intestinal mucosa. *J Pharm Sci* 57:795–798, 1968.
39. N Hashimoto, H Ohyanagi. Systemic factors are trophic in Thiry-Vella loop in the absence of luminal contents. *Hepatogastroenterology* 49:428–431, 2002.
40. RD Evans, MJ Bennett, D Hauton. Perfused heart studies to investigate lipid metabolism. *Biochem Soc Trans* 28:113–120, 2000.
41. HM Mehendale, LS Angevine, Y Ohmiya. The isolated perfused lung: a critical evaluation. *Toxicology* 21:1–36, 1981.
42. RW Niemeier. The isolated perfused lung. *Environ Health Perspect* 56:35–41, 1984.
43. A Zarzuelo, A Sanchez-Navarro, FG Lopez, JM Lanao. A review of the isolated kidney as an experimental model for pharmacokinetic studies. *Methods Find Exp Clin Pharmacol* 22:757–763, 2000.
44. KL Brouwer, RG Thurman. Isolated perfused liver. *Pharm Biotechnol* 8:161–192, 1996.
45. JW Porteous. Intestinal metabolism. *Environ Health Perspect* 33:25–35, 1979.
46. H Bohets, P Annaert, G Mannens, L van Beijsterveldt, K Anciaux, P Verboven, W Meuldermans, K Lavrijsen. Strategies for Absorption screening in drug discovery and development. *Curr Top Med Chem* 1:367–383, 2001.
47. HF Woods, MB Youdim. The isolated perfused rat brain preparation: a critical assessment. *Essays Neurochem Neuropharmacol* 2:49–69, 1977.
48. Z Hu, PG Wells. Modulation of benzo[a]pyrene bioactivation by glucuronidation in lymphocytes and hepatic microsomes from rats with a hereditary deficiency in bilirubin UDP-glucuronosyltransferase. *Toxicol Appl Pharmacol* 127:306–313, 1994.
49. Gonzalez FJ. The use of gene knockout mice to unravel the mechanisms of toxicity and chemical carcinogenesis. *Toxicol Lett* 120:199–208, 2001.

50. F Gonzalez, S Kimura. Study of P450 function using gene knockout and transgenic mice. *Arch Biochem Biophys* 409:153–158, 2003.
51. JH Lin, M Yamazaki. Role of P-glycoprotein in pharmacokinetics: clinical implications. *Clin Pharmacokinet* 42:59–98, 2003.
52. SA Roberts. High-throughput screening approaches for investigating drug metabolism and pharmacokinetics. *Xenobiotica* 31:557–589, 2001.
53. G Severin. A novel column/buffer combination for chromatographic analysis of basic drugs in simulated gastrointestinal fluids. *J Pharm Sci* 75:211–214, 1986.
54. P Artursson, K Palm, K Luthman. Caco-2 monolayers in experimental and theoretical predictions of drug transport. *Adv Drug Delivery Syst* 22:67–84, 1996.
55. JD Irvine, L Takahashi, K Lockhart, J Cheong, JW Tolan, HE Selick, JR Grove. MDCK (Madin–Darby canine kidney) cells: a tool for membrane permeability screening. *J Pharm Sci* 88:28–33, 1999.
56. WS Putnam, S Ramanathan, L Pan, LH Takahashi, LZ Benet. Functional characterization of monocarboxylic acid, large neutral amino acid, bile acid and peptide transporters, and P-glycoprotein in MDCK and Caco-2 cells. *J Pharm Sci* 91:2622–2635, 2002.
57. FR Luo, PV Paranjpe, A Guo, E Rubin, P Sinko. Intestinal transport of irinotecan in Caco-2 cells and MDCK cells overexpressing efflux transporters Pgp, cMOAT, and MRP1. *Drug Metab Dispos* 30:763–770, 2002.
58. A Avdeef, M Strafford, E Block, MP Balogh, W Chambliss, I Khan. Drug absorption *in vitro* model: filter-immobilized artificial membranes. 2. Studies of the permeability properties of lactones in *Piper methysticum* Forst. *Eur J Pharm Sci* 14:271–280, 2001.
59. IJ Hidalgo. Assessing the absorption of new pharmaceuticals. *Curr Top Med Chem* 1:385–401, 2001.
60. I Megard, A Garrigues, S Orlowski, S Jorajuria, P Clayette, E Ezan, A Mabondzo. A co-culture-based model of human blood-brain barrier: application to active transport of Indinavir and *in vivo–in vitro* correlation. *Brain Res* 927:153–167, 2002.
61. I Kariv, H Cao, KR Oldenburg. Development of a high throughput equilibrium dialysis method. *J Pharm Sci* 90:580–587, 2001.
62. LN Sansom, AM Evans. What is the true clinical significance of plasma protein binding displacement interactions? *Drug Safety* 12:227–233, 1995.
63. SA Wrighton, M Vandenbranden, JC Stevens, LA Shipley, BJ Ring, AE Rettie, JR Cashman. *In vitro* methods for assessing human hepatic drug metabolism: their use in drug development. *Drug Metab Rev* 25:453–484, 1993.
64. K Venkatakrishnan, LL Von Moltke, DJ Greenblatt. Human drug metabolism and the cytochromes P450: application and relevance of *in vitro* models. *J Clin Pharm* 41:1149–1179, 2001.
65. MN Berry, AR Grivell, MB Grivell, JW Phillips. Isolated hepatocytes: past, present and future. *Cell Biol Toxicol* 13:223–233, 1997.
66. D Runge, C Kohler, VE Kostrubsky, D Jager, T Lehmann, DM Runge, U May, DB Stolz, SC Strom, WE Fleig, GK Michalopoulos. Induction of cytochrome P450 (CYP)1A1, CYP1A2, and CYP3A4 but not of CYP2C9, CYP2C19, multidrug resistance (MDR-1) and multidrug resistance associated protein (MRP-1) by prototypical inducers in human hepatocytes. *Biochem Biophys Res Commun* 273:333–341, 2000.

67. RJ Edwards, RJ Price, PS Watts, AB Renwick, JM Tredger, AR Boobis, BG Lake. Induction of cytochrome P450 enzymes in cultured precision-cut human liver slices. *Drug Metab Dispos* 31:282–288, 2003.
68. PJ Eddershaw, M Dickins. Advances in drug metabolism screening. *Pharm Sci Technol Today* 3:13–19, 1999.
69. AP Beresford. CYP1A1: friend or foe? *Drug Metab Rev* 25, 503–517, 1993.
70. R Curi-Pedrosa, M Daujat, L Pichard, JC Ourlin, P Clair, L Gervot, P Lesca, J Domergue, H Joyeux, G Fourtanier. Omeprazole and Lansoprazole are mixed inducers of CYP1A and CYP3A in human hepatocytes in primary culture. *J Pharmacol Exp Ther* 269:384–392, 1994.
71. AP Beresford, WJ Ellis, J Ayrton, MA Johnson, DFV Lewis. Cytochrome P450 1A (CYP1A) induction in rat and man by the benzodioxino derivative, Fluparoxan. *Xenobiotica* 27:159–173, 1997.
72. J Pan, Q Xiang, AB Renwick, RJ Price, SE Ball, J Kao, JA Scatina, BG Lake. Use of real-time reverse transcription–polymerase chain reaction method to study the induction of CYP1A, CYP1A2 and CYP4A forms in precision-cut rat liver slices. *Xenobiotica* 32:739–747, 2002.
73. JL Staudinger, A Madan, KM Carol, A Parkinson. Regulation of drug transporter gene expression by nuclear receptors. *Drug Metab Dispos* 31:523–527, 2003.
74. G Luo, M Cunningham, S Kim, T Burn, J Lin, M Sinz, G Hamilton, C Rizzo, S Jolley, D Gilbert, A Downey, D Mudra, R Graham, K Carroll, J Xie, A Madan, A Parkinson, D Christ, B Selling, E LeCluyse, LS Gan. CYP3A4 induction by drugs: correlation between a Pregnane-X receptor reporter gene assay and CYP3A4 expression in human hepatocytes. *Drug Metab Dispos* 30:795–804, 2002.
75. PD Worboys, DJ Carlile. Implications and consequences of enzyme induction on preclinical and clinical development. *Xenobiotica* 31:539–556, 2001.
76. F De Clerck, A van de Water, J D'Aubioul, HR Lu, K van Rossem, A Hermans, K van Ammel. *In vivo* measurement of QT prolongation, dispersion and arrhythmogenesis: application to the preclinical cardiovascular safety pharmacology of a new chemical entity. *Fundam Clin Pharmacol* 16:125–140, 2002.
77. F De Ponti, E Poluzzi, A Cavalli, M Recanatini, N Montanaro. Safety of nonantiarrhythmic drugs that prolong the QT interval or induce Torsade de Pointes: an overview. *Drug Safety* 25:263–286, 2002.
78. B Surawicz. Torsades de Pointes: unanswered questions. *J Nippon Med School* 69:218–223, 2002.
79. DC Blanchard, G Griebel, RJ Blanchard. The mouse defense test battery: pharmacological and behavioral assays for anxiety and panic. *Eur J Pharm* 463:97–116, 2003.
80. DA Groneberg, C Grosse-Siestrup, A Fischer. *In vitro* models to study hepatotoxicity. *Toxicol Pathol* 30:394–399, 2002.
81. BL Pool, D Schmahl. What is new in mutagenicity and carcinogenicity: status of short-term assay systems as tools in genetic toxicology and carcinogenesis. *Pathol Res Pract* 182:704–712, 1987.
82. D Purves, C Harvey, D Tweats, CE Lumley. Genotoxicity testing: current practices and strategies used by the pharmaceutical industry. *Mutagenesis* 10:297–312, 1995.
83. PJ Kramer. Genetic toxicology. *J Pharm Pharmacol* 50:395–405, 1998.
84. K Muller-Tegethoff, B Kersten, P Kasper, L Muller. Applications of the *in vitro* rat hepatocyte micronucleus assay in genetic toxicology testing. *Mutat Res* 392:125–138, 1997.

85. K Mortelmans, E Zeiger. The Ames Salmonella/microsome mutagenicity assay. *Mutat Res* 455:29–60, 2000.
86. JH Weisburger. Antimutagenesis and anticarcinogenesis, from the past to the future. *Mutat Res* 480/481:23–35, 2001.
87. P Moller, LE Knudsen, S Loft, H Wallin. The comet assay as a rapid test in biomonitoring occupational exposure to DNA-damaging agents and effect of confounding factors. *Cancer Epidemiol Biomarkers Prevent* 9:1005–1015, 2000.
88. S Pavanello, E Clonfero. Biomarkers of genotoxic risk and metabolic polymorphism. *La Medicina Lavaro* 91:431–469, 2000.
89. RJ Albertini. HPRT mutations in humans: biomarkers for mechanistic studies. *Mutat Res* 489:1–16, 2001.
90. AR Collins. The comet assay. Principles, applications, and limitations. *Methods Mol Biol* 203:163–177, 2002.
91. C Schleger, N Krebsfaenger, A Kalkuhl, R Bader, T Singer. Innovative cell culture methods in drug development. *Alternativen Tierexperimenten* 18:5–8, 2001.
92. PJ Bugelski, U Atif, S Molton, I Toeg, PG Lord, DG Morgan. A strategy for high throughput cytotoxicity screening in pharmaceutical toxicology. *Pharm Res* 17:1265–1272, 2000.
93. SM Evans, A Casartelli, E Herreros, DT Minnick, C Day, E George, C Westmoreland. Development of a high throughput *in vitro* toxicity screen predictive of high acute *in vivo* toxic potential. *Toxicol In Vitro* 15:579–584, 2001.
94. S Kohler, S Belkin, RD Schmid. Reporter gene bioassays in environmental analysis. *Fresenius' J Anal Chem* 366:769–779, 2000.
95. VA Baker. Endocrine disrupters: testing strategies to assess human hazard. *Toxicol In Vitro* 15:413–419, 2001.
96. D Marzin. New approaches to estimating the mutagenic potential of chemicals. *Cell Biol Toxicol* 15:359–365, 1999.
97. WD Pennie, JD Tugwood, GJ Oliver, I Kimber. *Toxicol Sci* 54:277–283, 2000.
98. HK Hamadeh, RP Amin, RS Paules, CA Afshari. An overview of toxicogenomics. *Curr Issues Mol Biol* 4:45–56, 2002.
99. JF Waring, DN Halbert. The promise of toxicogenomics. *Curr Opin Mol Ther* 4:229–235, 2002.
100. RP Amin, HK Hamadeh, PR Bushel, L Bennett, CA Afshari, RS Paules. Genomic interrogation of mechanism(s) underlying cellular responses to toxicants. *Toxicology* 181/182:555–563, 2002.
101. MJ Aardema, JT MacGregor. Toxicology and genetic toxicology in the new era of "toxicogenomics": impact of "-omics" technologies. *Mutat Res* 499:13–35, 2002.
102. LR Bandara, S Kennedy. Toxicoproteomics: a new preclinical tool. *Drug Discov Today* 7:411–418, 2002.
103. E Holmes, JK Nicholson, G Tranter. Metabonomic characterization of genetic variations in toxicological and metabolic responses using probabilistic neural networks. *Chem Res Toxicol* 14:182–191, 2001.
104. JP Shockcor, E Holmes. Metabonomic applications in toxicity screening and disease diagnosis. *Curr Top Med Chem* 2:35–51, 2002.
105. MR Fielden, JB Mathews, KC Fertuck, RG Halgren, TR Zacharewski. *In silico* approaches to mechanistic and predictive toxicology: an introduction to bioinformatics for toxicologists. *Crit Rev Toxicol* 32:67–112, 2002.

106. CJ Manly, S Louise-May, JD Hammer. The impact of informatics and computational chemistry on synthesis and screening. *Drug Discov Today* 6:1101–1110, 2001.
107. DK Agrafiotis, VS Lobanov, FR Salemme. Combinatorial informatics in the postgenomic era. *Nat Rev* 1:337–346, 2002.
108. J Xu, A Hagler. Chemoinformatics and drug discovery. *Molecules* 7:566–600, 2002.
109. MG Bures, YC Martin. Computational methods in molecular diversity and combinatorial chemistry. *Curr Opin Chem Biol* 2:376–380, 1998.
110. P Willett. Similarity and diversity in chemical libraries. *Curr Opin Biotechnol* 11:104–107, 2000.
111. DK Agrafiotis. Stochastic algorithms for maximizing molecular diversity. *J Chem Inf Comput Sci* 37:841–851, 1997.
112. RS Pearlman, KM Smith. Software for chemical diversity in the context of accelerated drug discovery. *Drugs Future* 23:885–895, 1998.
113. RS Pearlman. Novel Software Tools for Addressing Chemical Diversity. Network Science, http://www.netsci.org/Science/Combichem/feature08.html, 1996.
114. TC Kuhler, J Gottfries, TI Oprea, V Sherbukhin, P Svensson. Chemical information management in drug discovery: optimizing the computational and combinatorial chemistry interfaces. *J Mol Graph Model* 18:512–524, 2000.
115. EJ Martin, JM Blaney, MA Siani, DC Spellmeyer, AK Wong, WH Moos. Measuring diversity: experimental design of combinatorial libraries for drug discovery. *J Med Chem* 38:1431–1436, 1995.
116. EJ Martin, R Critchlow. Beyond mere diversity: tailoring combinatorial libraries for drug discovery. *J Comb Chem* 1:32–45, 1999.
117. RA Fecik, KE Frank, EJ Gentry, SR Menon, LA Mitscher, H Telikepalli. The search for orally active medications through combinatorial chemistry. *Med Res Rev* 18:149–185, 1998.
118. SJ Teague, AM Davis, PD Leeson, T Oprea. The design of leadlike combinatorial libraries. *Agnew Chem Int Ed* 38:3743–3748, 1999.
119. TI Oprea. Current trends in lead discovery: are we looking for the appropriate properties? *J Comput Aided Mol Des* 16:325–334, 2002.
120. WJ Eagan, WP Walters, MA Murko. Guiding molecules toward drug-likeness. *Curr Opin Drug Discov Dev* 5:540–549, 2002.
121. MM Hann, AR Leach, G Harper. Molecular complexity and its impact on the probability of finding leads for drug discovery. *J Chem Inf Comput Sci* 41:856–864, 2001.
122. GM Rishton. Nonleadlikeness and leadlikeness in biochemical screening. *Drug Discov Today* 8:86–96, 2003.
123. I Muegge. Selection criteria for drug-like compounds. *Med Res Rev* 23:302–321, 2003.
124. R Todeschini, V Consonni, Eds. *Handbook of Molecular Descriptors*. Wienheim, Germany: Wiley-VCH, 2000.
125. A Ajay, WP Walters, MA Murcko. Can we learn to distinguish between "drug-like" and "nondrug-like" molecules? *J Med Chem* 41:3314–3324, 1998.
126. TI Oprea. Property distribution of drug-related chemical databases. *J Comput Aided Mol Des* 14:251–264, 2000.
127. JF Blake. Chemoinformatics: predicting the physicochemical properties of "drug-like" molecules. *Curr Opin Biotechnol* 11:104–107, 2000.
128. SD Pickett, IM McLay, DE Clark. Enhancing the hit-to-lead properties of lead optimization libraries. *J Chem Inf Comput Sci* 40:263–272, 2000.

129. RS Pearlman, KM Smith. Metric validation and the receptor-relevant subspace concept. *J Chem Inf Comput Sci* 39:28–35, 1999.
130. JG Topliss. Computer-aided drug design in industrial research: a management perspective. In *Computer-Aided Drug Design in Industrial Research*, EC Herrmann, R Franke, Eds. Berlin: Springer-Verlag, 1995, chap. 2.
131. C Hansch, A Leo, D Hoekman. *Exploring QSAR. Hydrophobic, Electronic and Steric Constants*. American Chemical Society, 1995.
132. MA Navia, PR Chaturvedi. Design principles for orally bioavailable drugs. *Drug Discov Today* 1:179–189, 1996.
133. CA Lipinski, F Lombardo, BW Dominy, PJ Feeney. Experimental and computational approaches to estimate solubility and permeability in drug discovery and development settings. *Adv Drug Deliv Rev* 23:3–25, 1997.
134. MC Wenlock, RP Austin, P Barton, AM Davis, PD Leeson. A comparison of physiochemical property profiles of development and marketed oral drugs. *J Med Chem* 46:1250–1256, 2003.
135. YH Zhao, J Le, MH Abraham, A Hersey, PJ Eddershaw, CN Luscombe, D Boutina, G Beck, B Sherborne, I Cooper, JA Platts. Evaluation of human intestinal absorption data and subsequent derivation of a quantitative structure-activity relationship (QSAR) with the Abraham descriptors. *J Pharm Sci* 90:749–784, 2001.
136. P Crivori, G Cruciani, PA Carrupt, B Testa. Predicting blood-brain barrier permeation from three-dimensional molecular structure. *J Med Chem* 43:2204–2216, 2000.
137. KR Korzekwa, JP Jones, JR Gillete. Theoretical studies on cytochrome P-450 mediated hydroxylation: a predictive model for hydrogen atom abstractions. *J Am Chem Soc* 1112:7042–7046, 1990.
138. SB Singh, LQ Shen, MJ Walker, RP Sheridan. A model for predicting likely sites of CYP3A4-mediated metabolism on drug-like molecules. *J Med Chem* 46:1330–1336, 2003.
139. S Ekins, CL Waller, PW Swaan, G Cruciani, SA Wrighton, JH Wikel. Progress in predicting human ADME parameters *in silico*. *J Pharmacol Toxicol Methods* 44:251–272, 2001.
140. MP Beavers, X Chen. Structure-based combinatorial library design: methodologies and applications. *J Mol Graph Model* 20:463–468, 2002.
141. TG Davies, J Bentley, CE Arris, FT Boyle, NJ Curtin, JA Endicott, AE Gibson, BT Golding, RJ Griffin, IR Hardcastle, P Jewsbury, LN Johnson, V Mesguiche, DR Newell, ME Noble, JA Tucker, L Wang, HJ Whitfield. Structure-based design of a potent purine-based cyclin-dependent kinase inhibitor. *Nat Struct Biol* 9:745–749, 2002.
142. N Green, PN Judson, JJ Langowski, CA Marchant. Knowledge-based expert systems for toxicity and metabolism prediction: DEREK, STAR and METEOR. *SAR QSAR Environ Res* 10:299–314, 1999.
143. G Klopman. Artificial intelligence approach to structure-activity studies. Computer automated structure evaluation of biological activity of organic molecules. *J Am Chem Soc* 106:7315–7321, 1984.
144. D Zmuidinavicius, P Japertas, A Petrauskas, R Didziapetris. Progress in toxinformatics: the challenge of predicting acute toxicity. *Curr Top Med Chem* 3:1301–1314, 2003.
145. PA Bacha, HS Gruver, BK Den Hartog, SY Tamura, RF Nutt. Rule extraction from a mutagenicity data set using adaptively grown phylogenetic-like trees. *J Chem Inf Comput Sci* 42:1104–1111, 2002.

146. P Gedeck, P Willet. Visual and computational analysis of structure-activity relationships in high-throughput screening data. *Curr Opin Struct Biol* 5:389–395, 2001.
147. Diva®, Accelrys, Inc. Accelrys is a wholly owned subsidiary of Pharmacopeia, Inc.
148. Spotfire®, Spotfire, Inc., Somerville, MA.
149. Omniviz, OmniViz, Inc., Maynard, MA.
150. G Roberts, GJ Myatt, WP Johnson, KP Cross, PE Blower, Jr. Leadscope: software for exploring large sets of screening data. *J Chem Inf Comput Sci* 40:1302–1314, 2000.
151. JM McKenna, F Halley, JE Souness, IM McLay, SD Pickett, AJ Collis, K Page, I Ahmed. An algorithm-directed two-component library synthesized via solid-phase methodology yielding potent and orally bioavailable p38 MAP kinase inhibitors. *J Med Chem* 45:2173–2184, 2002.
152. G Bravi, DVS Green, MM Hann, AR Leach. PLUMS: a program for the rapid optimization of focused libraries. *J Chem Inf Comput Sci* 40:1441–1448, 2000.
153. VJ Gillet, P Willet, J Bradshaw, DVS Green. Selecting combinatorial libraries to optimize diversity and physical properties. *J Chem Inf Comput Sci* 39:169–177, 1999.
154. W Zheng, ST Hung, JT Saunders, GL Seibel. PICCOLO: a tool for combinatorial library design via multicriterion optimization. *Pac Symp Biocomput* 588–599, 2000.
155. RD Brown, M Hassan, M Waldman. Combinatorial library design for diversity, cost efficiency, and drug-like character. *J Mol Graph Model* 18:427–437, 2000.
156. J Shimada, S Ekins, C Elkin, EI Shakhnovich, JP Wery. Integrating computer-based *de novo* drug design and multidimensional filtering for desirable drugs. *Targets* 1:196–205, 2002.
157. SL Gallion, L Hardy, A Sheldon. Tilting toward targets: biased compound sets. *Curr Drug Discov* 1:25–27, 2002.
158. AB Beresford, M Tarbit, K Jessing. Application of the Camitro Predictive ADME/Tox Platform for Lead Optimisation. Paper presented at International Business Conferences: 4th Annual Conference on Predictive and Emerging Technologies for Lead Selection and Optimization, November 2001.
159. SL Gallion. Biased Library Design and Parallel Optimization of Ion Channel Modulators. Paper presented at International Business Conferences: Advances in Structure- Guided Drug Discovery, San Diego, CA, November 2002.
160. VJ Gillet, W Khatib, P Willett, PJ Fleming, DVS Green. Combinatorial library design using a multiobjective genetic algorithm. *J Chem Inf Comput Sci* 42:375–385, 2002.
161. S Ekins, B Boulanger, PW Swaan, MAZ Hupcey. Toward a new age of virtual ADME/TOX and multidimensional drug discovery. *J Comput Aided Mol Des* 16:381–401, 2002.
162. Admensa Interactive, Inpharmatica Ltd., Cambridge, U.K.

chapter 7

Knowledge Management

Beverly Buckta

Contents

7.1 What Is Knowledge Management?

There is a large demand in organizations for people who are interested in understanding information and knowledge management within a corporate environment. Information and knowledge management (KM) are burgeoning fields and there is a high demand in organizations for people with information and knowledge management proficiency and practice. Companies realize the need to access the knowledge base of their employees contained within them as well as in the innumerable documents and files they generate each day. A prerequisite of knowledge management is the foundational capability of finding, accessing and exploiting information on an enterprise-wide scale. People with the ability to diagnose informational needs, retrieve relevant information, and repackage that information in the most appropriate manner are in high demand. Organizations recognize that improved efficiency and effectiveness can be gained by improving knowledge accessibility wherever it might exist.

Knowledge management means many things to many people. The extreme responses to KM, as reflected in the literature and in practice, are skepticism, on the one hand, to "the next silver bullet" on the other. Part of the dilemma is that many of the definitions in the KM literature are full of jargon. Some examples of knowledge management definitions include:

- Knowledge management is the collection of processes that govern the creation, dissemination, and utilization of knowledge.
- Knowledge management processes consist of the creation, collection, interpretation, and storage of, as well as interaction with, data.
- Knowledge management is the discipline that promotes an integrated approach to identifying, managing, and sharing all of the enterprise's information assets, including databases, documents, policies, and procedures, as well as unarticulated expertise and experience resident to individual workers.
- Knowledge management is the management of the organization toward the continuous renewal of the organizational knowledge base—this means, for example, the creation of supportive organizational structures, facilitation of organizational members, and putting into place IT instruments with emphasis on teamwork and diffusion of knowledge (as, for example, groupware).
- A knowledge flow process that reaches well beyond having excellent data/information storage and retrieval to embrace retrieval, creation, capture, use, and reuse of knowledge and information for innovation.
- The broad process of locating, organizing, transferring, capturing, and using the information and expertise within an organization. Four key enablers support the overall knowledge management process: people, information, technology, and process.
- Knowledge management caters to the critical issues of organizational adaption survival, and competence in the face of increasingly

discontinuous environmental change. Essentially, it embodies organizational processes that seek a synergistic combination of data and information processing capacity of information technologies, and the creative and innovative capacity of human beings.

- During the first 6 months of 2000, a study was sponsored by SCRIP (17), where 120 companies were examined: 42 claimed relevant activities and 7 used a formal KM label. "The common objective of activities relevant to our research is improved business performance, through the creation, sharing, and utilization of knowledge."

Although the term *knowledge management* can have many meanings, the concept is based on common sense and good management practice. Knowledge management has come about because companies, especially larger ones, need to make more efficient, effective, and fluid decisions to be competitive. In addition, as electronic information (eSpace) has grown, companies are looking at their processes to identify processes that are commodities and which provide competitive advantage. Companies have enormous investments in information technology and proving management exploitation of these investments are key. Knowledge management is multidimensional and involves holistic, systemic thinking. This thinking revolves around life cycles, including the product life cycle and information life cycle. Individual, team or group, and enterprise needs related to the levels of data, information, and KM is an important aspect to address. The pharmaceutical value chain, and considerations about supporting the depth of processes or subprocesses and breadth of process integration is key in determining the best approach to data, information, or knowledge management. These are a few of the multiple dimensions which enable the flow of data, information, and ultimately knowledge in an organization.

Organizations are structured, but the flow of information and content for effective decision-making is not structured. In fact, the innovative thinking of people is usually captured in unstructured information containers (documents) and these documents get stored all over the organization. People's most innovative thinking is often unstructured. So how do companies take advantage of and benefit from the thinking of their employees and the information and content they generate, while compensating for the rigid, but necessary, organizational structure?

KM broadly requires efficient and effective information management (IM) and an enterprise (umbrella) framework to support people, information, processes, and technology. Information management and knowledge management are not synonymous, but IM is essential to the realization of KM. There is a tremendous amount of electronic information generated in companies each day as evident by the amount of companies now investigating electronic (e) storage alternatives. With so much e-information, the meaning of information theory and information management has taken on new meanings.

There is also pure information theory, which was initially developed for application as telecommunications, where it is both possible and feasible to compute the amount of information that can be transmitted over a wire or radio band. For determining channel capacity, it has indeed been of significant benefit. However, its utility when applied to other disciplines has been written with the many claims regarding the importance of information theory and its applicability to the theory of business organizations. There have been varying degrees of value, and the attempts to apply it to the business sector have been equally satisfying and disappointing (15). In recent years, some have endeavored to apply the formal theory to the fields of experimental psychology, sociology, decision-making, accounting, and many other diverse situations. In R&D, for example, informatics combines computational biology and chemistry, data modeling and integration, data mining, and visualization, such as mathematical modeling and simulation or image processing and analysis, with information theory and semantics. Information management is the broader spectrum of supporting the flow of information in support of business processes and managing information as corporate assets from the "cradle to grave."

In order to realize IM, there needs to be governance, principles, behavioral incentives, cultural norms, enabling processes, and an enterprise framework. The enterprise framework includes management of information life cycles, an IM architectural framework, and enterprise information services. The enterprise information services are those information solutions (the processes, IM, and technology) such as eCollaboration (particularly virtual teaming elements), search (in support of locating and accessing information and expertise), document management, content management e-mail, and other fundamental activities that can benefit from a central approach. eCollaboration, for example, is similar to e-mail, as we could not imagine going back to the days when the enterprise had more than one e-mail solution. Why is eCollaboration any different? These information services often are not considered infrastructure within companies because the amount of decisions regarding information that need to be made with each one is more than IT wants. Since these information services benefit the whole enterprise, no one business area is typically willing to pay for everyone's benefit. Therefore, these information services need a home. Perhaps information systems (IS) in many companies can step up to the plate with the concentration on where information and knowledge within the enterprise have potential to be harnessed.

7.2 *Why Is Knowledge Management Important?*

There are many examples of instances where companies embraced KM. The reasons include the changing business environment, and tight economic times resulting in the need for companies to become more efficient and effective by better leveraging the intellectual capital within the company. Davis and Meyer (5) describe "economy" as the way people use resources to fulfill their desires. "The specific ways they do this have changed several times through history, and are shifting yet again—this time driven by three forces—Connectivity, Speed, and the growth of intangible value." Knowledge management is an integral component in the evolution from industrial era focus on products and services to an economic, information, and emotional exchange.

"The task of making knowledge work productive will be the great challenge of the century, just as to make manual work productive was the great management task of the last century." Although Peter Drucker (7) wrote this in 1974, it still holds true today. When we speak of connected individuals and their knowledge as an enterprise's key organizing unit we are reminded of the importance of knowledge management. Pharmaceutical companies are charged with managing both products and services, and also with managing the collective knowledge of their employees and their information, which is essential for pharmaceutical products. Constancy does not exist in the information age: we see vast transformations in the nature of work and continuous change in the work environment—hallmarks of the knowledge workplace (Table 7.1).

Companies today are also challenged with functioning as extended organizations. Downey et al. (6) describe the horizontal organization. Manufacturing across the industrial sector is going *horizontal,* meaning that vertically integrated supply chains are breaking apart into component layers dominated by horitzontal specialists, thus creating cost-effective combinations. This seismic shift is reshaping the future of R&D around the world. Organizations with a global orientation that can excel collaboration and financial operations can generate greater revenues and higher margins through this horizontal evolution. Effective knowledge management is critical to the horizontal organization, where information and knowledge need to flow throughout these combinations. With every R&D investment now being scrutinized, effective knowledge management is needed to help evolve to a horizontal future.

Management in the industrial sector has matured, based on evolving practices over the past few decades. Perhaps knowledge management is an early form of new management evolving to sustain innovation in new ways. Knowledge sharing is a catalyst for creativity and innovation because it provides a means for innovative ideas to be captured, shared, and tested. That is why there is a growing appreciation for KM involvement earlier in the R&D scientific as well as commercial assessment (2).

Table 7.1 Industrial Age vs. Knowledge Workplace Maxims

Industrial Age Maxim	Knowledge Workplace Maxim
Preserve the status quo	Promote change as creative and imperative
Resist organizational movement	Encourage organizational movement
Engender rugged individualism	Inspire connected teamwork
Command and control workers	Lead by facilitating utmost creativity
Organize for uniformity	Respect worker individuality
Divide and control	Collaborate and innovate
People need jobs	Value each contribution

Adapted from D Wesley, *Information Outlook* 2000, 35–40.

KM has been shown to improve success of investment decisions, decrease product-to-market times, and develop world-class customer intimacy and satisfaction (17). Actual time spent on core work and competencies (as opposed to searching, validating information, building models, and supporting tools) has been shown to increase from 25 to 75%. Model reuse has been shown to increase from an average of less than 2 times to more than 10 times per model. Abandoned projects dropped from 25 to 1%. A number of pharmaceutical and biotechnology companies have reported significantly improved R&D initiatives, particularly involving data visualization and decision support functionality (16). KM, as one tool that can be used, is expected to impact R&D speed because of the quicker access and flow of the most relevant information, faster decision making, and broader sharing of best practices (1).

7.3 *Turning Bytes into Drugs: Evolving Demand*

The pharmaceutical industry is not shielded from the dilemma of needing knowledge management. And with the evolving demands on the pharmaceutical industry, the need to produce more with less is critical. Pharmaceutical companies tend to have more experience with the management of structured data. And since there is ever-increasing unstructured information being created, captured, and needed, there is also an ever-increasing challenge in attempting to deal with the information and knowledge.

Adopting a strategic orientation results in IS/IT systems alignment to business strategy, requiring coopertation and coordination between IS managers and business unit managers. The need for flexible and adaptable systems becomes evident as business process reengineering, restructuring, shifting alliances, new competitors, globalization, legacy system migration, and new technology adoption, are only a few of the economic, organizational, and technological forces driving change and reshaping organizations. These changes are impacting the whole of the pharmaceutical value chain that supports the product life cycle and is placing more demand on the need for knowledge management. Companies have successfully supported the business processes related to the pharmaceutical product life cycle in the vertical style. In other words, there are examples of knowledge management solutions for individual parts (and subparts) of the pharmaceutical value chain. The challenge now is how to support business processes, information, and knowledge management horizontally across the value chain and extend them to external companies, CROs (clinical research organizations), and other partners who are also supporting that value chain.

Some examples of good vertical knowledge management, especially in areas with the need to deal with huge volumes of information, are in Discovery. Discovery deals with large volumes of detailed scientific information requiring data mining and analysis. Those in Clinical experience transactional information for trial execution. Those in Drug Safety deal with patient and drug-level summarization. Medical Affairs work with external partner

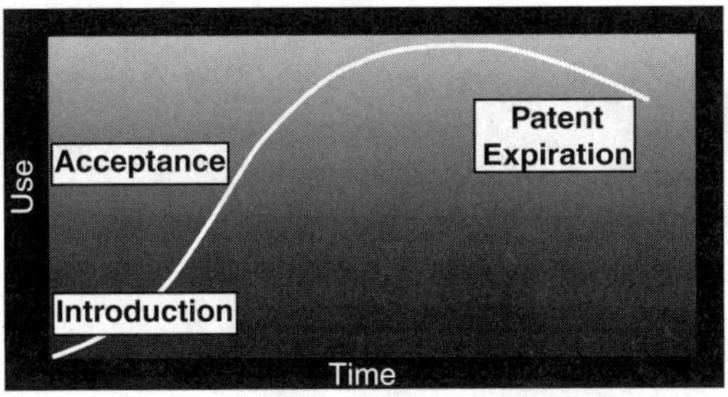

Figure 7.1 Pharmaceutical product life cycle.

information, while Sales and Marketing need to use information for commercial interests. The pharmaceutical company that best harnesses and makes use of the information and knowledge across these functions is the company that will out perform its competitors.

Considering the pharmaceutical product life cycle (Figure 7.1), a subsection of the above curve comprises processes within discovery and development in support of product launch. Furthermore, each process within discovery and development can be further defined as a process in support of a broad registration. Each of the processes and subprocesses needs information and knowledge *flows*. The more awkward the flows, the more awkward the processes, especially when the processes support geographically-dispersed teams who are working in a constantly changing market that is customer-driven, under pressure, global-connected, supported by rapidly changing information technologies, and who require management techniques that support efficiency, effectiveness, and productivity, as well as innovation, learning, and serendipity. A large number of factors can cause the product life cycle curve to shift. The information life cycle (and a knowledge life cycle) can be applied perpendicular to any one point on any of the process flows (analogy is a slide rule). The company who embeds the integration of information and knowledge into these and other business processes, as a typical way of working, will be a winning company. Information and knowledge management need to be embedded as business practices in themselves, which will depend on cultural change and very effective leadership (17). Also, because these processes and subprocesses need to be viewed as a holistic system, KM needs to be addressed at all levels, including enterprise-wide in parallel to local, more departmental levels. The subprocesses need to be supported by their broader processes. Figure 7.2 (the pharmaceutical value chain) oversimplifies the nature of the pharmaceutical processes, especially for R&D, because it suggests a linear, as opposed to an iterative, approach.

So, as we look across the value chain, we can discuss examples of good vertical knowledge management. And we can, in fact, illuminate the areas that should be leveraged in the company more as information and knowledge management is "kicked up a notch" for a more holistic, horizontal balance.

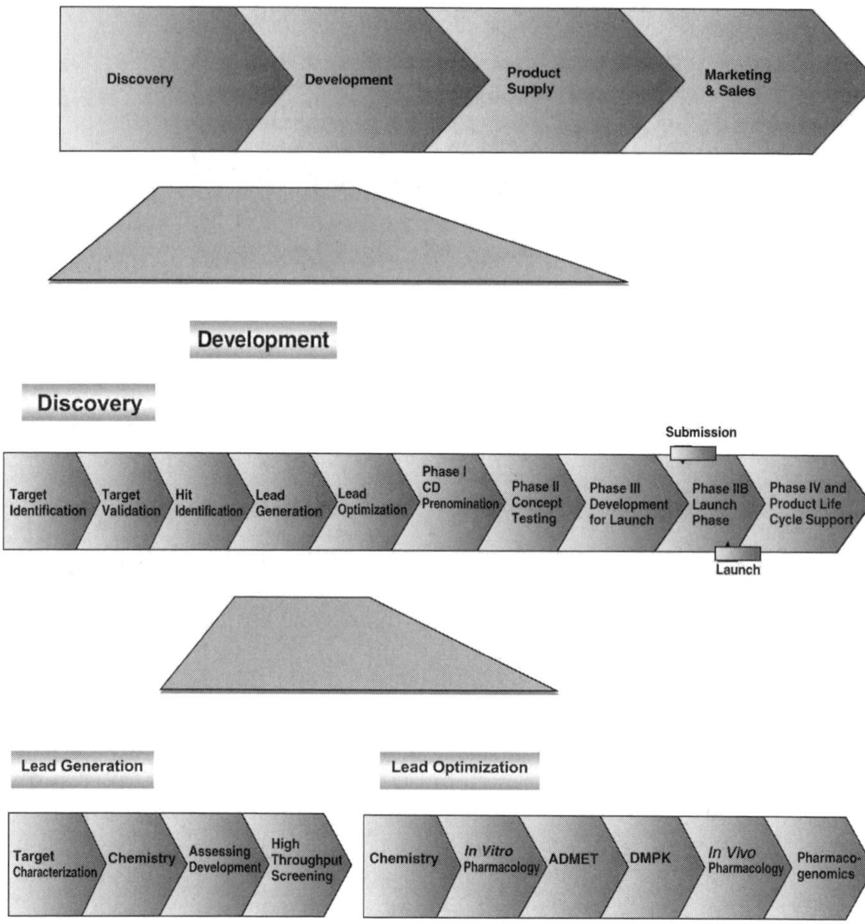

Figure 7.2 Pharmaceutical value chain with expanding selection and examples of business processes and subprocesses.

7.3.1 *Volumes of Detail for Data Mining/Analysis (Discovery)*

Drug discovery processes and technologies can create value by transforming vast amounts of data into knowledge that is used to create products that improve or save lives. The creation and volume of that data in the drug area is rapidly growing. In fact, its growth is at a faster rate than researchers are able to manage. The use of KM principles and tools is becoming crucial to reduce development times and costs and improve the overall success rate of testing new compounds (16).

Collaboration is key in Discovery due to the nature of the work. Improved internal and external collaboration with a single access point is critical in discovery, as well as the ability to search. Flexible access to a variety of compound libraries is necessary. The interaction between people and their ideas is core to eCollaboration (and broadly to KM), in particular, by supporting a more innovative workflow and virtual teaming. Virtual teaming "spaces" can be used for research, product planning, drug discovery, and regulatory processes. Facilitating and easing the creation and use of information is going to yield knowledge in those spaces without any additional conscious effort. Such use of a collaborative space may include resource management and alliance management. Alliances are an area where inter-organizational knowledge management is crucial for the growth and survival of firms. Knowing where to procure and access cutting-edge technology of other firms through alliance formation requires the need to make decisions across a research and knowledge-intensive industry. Mitsuhashi (12) writes, "Moreover, since drug discovery activities require a large financial investment and long-term commitment, the issue of selection uncertainty is more manifest in forming biopharmaceutical alliances." Regarding expertise alone, a wide variety of professionals need to know information, including: scientists who investigate and assess the competence of prospective partners; attorneys for intellectual property issues; finance and marketing professionals for cost–benefit analyses and estimating the market; manufacturing experts, clinical trial experts, and physicians to examine the potential medical applications of the products; experts in health insurance issues; and business development experts to assess proposed alliances and partnerships. Moreover, much of the scientific and technical know-how is not always explicit, but it is tacit. In order for upstream drug discovery alliances to be most effective, higher coordination and more frequent interactions must occur between scientists across organizational boundaries. Each of these experts also has several places to go for information with time pressures. So more coordination involves integration of content within individualized/ enterprise portals and integrated collaboration functionality.

Descriptions of R&D often depict a single industry class dominated by explicit knowledge. There is explicit knowledge in the form of patents, blueprints, diagrams, structures, and so forth, but this depiction often limits the ability to effectively manage the business processes across various stages of a science life cycle. The scientific knowledge base is well codified where there have been traditional forms of explicit knowledge (that is, product patents), but less so in developing fields (such as biotechnology). The knowledge that exists in product development differs across more traditional and developing industries. Cardinal et al. (4) have compared and contrasted the knowledge- and resource-based requirements of developed and developing science industries and demonstrated the link to competitive advantage. Knowledge may range from highly tacit (organizationally embedded, low codifiability) to easily codifiable. It is also equally important for knowledge barriers to protect knowledge-based resources from competitors. Most business processes involve a blend to leverage past experiences and new organization learning for value-added activities. This presents the challenge of extending what has been done well traditionally in terms of dealing with explicit knowledge and structured information and integrating unstructured and implicit knowledge. Not only is the competitive advantage in doing this integration well vertically within a business process, but also horizontally throughout processes and the extended enterprise. The knowledge and information flow from upstream activities (R&D) through production to downstream activities (marketing and sales). "R&D is useless unless it can be integrated into a bigger picture that helps you convert leads to drug candidates... The difficulty for the CIO is getting the team to work between the various disciplines needed to integrate the information in order to create knowledge" (14). There is huge change management that must take place before the infrastructure can be most effective.

7.3.2 *Transactional Information for Trial Execution (Clinical)*

In drug development, the accuracy of data, extended reach, and the speed of data collection and consolidation is key to developing candidate compounds. Cardinal et al. write that often times the new product development team has to "feel their way through the process" (4). The reason for this is that there is most often little existing explicit knowledge to draw upon to design the product or process. Without a mature knowledge base to draw upon, processes need to be highly interactive and concurrent. The process needs to be codeveloped with the product. Product development processes consist of iterations of upstream and downstream activities. For organizations that are involved in networks and alliances, processes need to be tightly coupled instead of separated out in stages to be successful.

Data modeling and quality is key here for information flow and integration. Applications are accessed for clinical trial management. Pharmaceutical companies work with contract research organizations to manage clinical trials for a selected subset of their new drug portfolio. More knowledge management strategies are helping companies address R&D issues, such as the need for speed, adequate yield, and rapid, shifting markets.

Faster query resolution and improved regulatory compliance is key in taking the emerging dossier consolidation through the regulatory approval cycle toward a marketable drug. The common technical document (CTD) and other initiatives to standardize regulatory information processing are examples where there are cost reduction and accelerated time-to-market opportunities.

Internal KM and organizational compliance are also to benefit from emerging formats such as the electronic version of the CTD (e-CTD). The electronic version enables viewing, navigation, searching, and assistance with report writing, as well as life cycle management. This first harmonization initiative standardizes structure and nomenclature of the drug marketing application and significantly increases the potential reuse of the authored content, finished and formatted forms/applications, and the publishing process on a worldwide basis (9). Depending on the regional definition of a product with different trade names, dosage forms, and strengths, a company with two products in

30 regions may have to create and track 1080 applications. The ability to reuse content when creating and maintaining global marketing applications is critical. Interestingly, however, the average company reuses only 10% of content in global submissions (9). So there is plenty of room for improvement.

7.3.3 Patient and Drug-Level Summarization (Drug Safety)

Drug Safety involves data collection and analyses is and is a subprocess of Drug Development. Effective data models are essential as data across databases need to be searched for specific linkages of data to identify statistical patterns. Numerous types of studies are conducted after drug approval including clinical studies, line extensions such as package sizes or dosage forms, changes in indications, mechanism of action studies, and postmarketing surveillance studies. Information on drug safety comes from human studies, patient experience, and toxicological studies. Huge volumes of data need to be sifted through, retained, and searchable with context.

Certain information has an information value that warrants an extended information life cycle. Data mining for structured and unstructured information is necessary for summarizing data into information, for example, deriving product labeling. Following case reports, formal studies, and other published information reported in the medical literature requires working with information horizontally across processes. Thus, good data, information, and knowledge management is crucial for the flow of information.

7.3.4 External Information Access and Supply with External Organizations (Operations)

Savings can be achieved with efficient Information and Knowledge Management. Small efficiencies in KM can result in huge impacts on avoiding production shutdowns and cutting inventories in half. Production failure rates can be impacted with good information and knowledge flow, among many other factors. Supplier score carding, forecasting and inventory collaboration design, and ECO management can streamline the iterative transition from product development. Of course, effective data management here, will most impact specific subprocesses in Operations, such as inventory management. Information management will most impact broader processes, through vertical integration, in support of Just-In-Time processes. A decision about what is produced, how, where, and how much and how well it is produced with quality control and quality assurance processes are some other examples of important vertical integration decisions and subprocesses. Most value would stem from enabling IM and KM across Operations in a more horizontal nature, which may be visible from a dashboard of sorts, which "pulls together" information across the subprocesses and ultimately across the broader pharmaceutical processes. Here, information logically presented to leadership, integrated from various primary sources and broad processes, gives decision-makers trusted information to make decisions about production alongside of, for example, Development and Marketing decisions.

Information may also be extended in reach outside of the company with, for example, external suppliers. The value in collaborating with raw material suppliers and multiple manufacturing plants introduces efficiencies to be had. This requires effective information and knowledge management to enable the unstructured and structured information exchanges.

7.3.5 Commercial Interests (Sales and Marketing)

If we consider where discovery and development integrates with operations and the marketing and sales business processes, the recycle value of information and knowledge management has huge potential. This is where the transition from information to knowledge management can create more value—tying business processes across major business functions that would normally be more vertical in nature. The flow of information that supports pricing decisions can take into account such items as the costs of raw supplies to make the drug, costs to manufacture and distribute, costs of active drugs used as controls in Phase II studies, and a drug's stage in its life cycle.

It is important to aggregate information and consolidate information and knowledge while creating content services (such as application or Web services), building on tangible value, then enhancing efficiencies across information stores. One approach would involve the move from disparate information sites to content management, to portal and content management, leading to collaborative networks to capture knowledge. Within marketing and sales, there is a great deal of information regarding marketing analytics and sales analytics. With the use of data models, the analytical information between marketing and sales can be integrated in a view. This would yield value in supporting decision-making across these data sets. By expanding these data sets with additional information and context, more comprehensive decision-making can take place and the knowledge management will be evident when there is someone new in a role and he or she can get up to speed in their responsibilities quickly since they have all the information they need.

Faster time to market can occur through effective pipeline management. KM can have an effect on efficiencies in client engagement, proposal development, and project portfolio management. Global and co-marketing benefit from KM. KM can facilitate government and financial services through information and knowledge flow across research and deal management. Contract management, policy development, program and project management, and emergency response management all generate and leverage information. The more the information can be accessed and reused, the more efficiencies and speed that can be gained. Measures may include savings in travel and increases in additional business with the same resources.

Throughout this all, greater flexibility will be offered through enterprise knowledge management with a common framework. The whole of the pharmaceutical value chain demands support research, development, sales, marketing, and corporate oversight of efforts within numerous separate operating companies. Rapidly expanding cross-collaboration efforts between operating companies as well as external partners are increasing in demand. The expected results are research teams and operations effectively sharing information on upcoming products with sales and marketing teams. The benefits of cross-organizational (horizontal) process integration through KM realize synergies.

7.4 More Efficient Access to More Integrated Information

Using an example of where KM has been accomplished well in a vertical fashion, let us discuss package inserts ([PIs], sometimes called product inserts) where there is a need for horizontal knowledge management. Package inserts contain information about a particular product, its indications, drug interactions, adverse effects, and other information about the product. Enterprise reuse—rendered in a portal, but even if horizontal solutions need enterprise architecture, framework, and enterprise portal. Pharmacogenomics, where DNA is studied in different people and their responses to a drug, is another area that will require marketing to understand if there are defined population subsets that will respond best to a particular product.

From a content publisher viewpoint, content collection and management are the essence of better managing the electronic knowledge spaces so that these spaces can permeate across organizational boundaries. Analyzing the reach and frequency of content, as well as the value and ease of content, starts to provide some methodological procedures on unbiasedly prioritizing content.

In order for knowledge management to be effective, companies need to be good at information management. The information life cycle needs to be applied at all points of the business processes. Without information management, integrated information does not occur, and therefore, it becomes more difficult to capture knowledge that can be trusted information. Information itself can be structured, unstructured, internal, external, validated, archived and stored, available for submission to regulatory agencies, domestic, global, partner work in progress, customer oriented, or at any stage of the information life cycle. Many of the outcomes lead to improved decision making. Not only does the information life cycle apply to each step of the pharmaceutical value chain, but the knowledge flow, at each business process or subprocess, happens with the idea creation, decision making, summarizing and analyzing, and new knowledge creation.

Armbrecht et al. (1) outlined a KM model that depicts the process above interacting with people in every stage of the cyclic process, from ideas through results. This interaction with information and knowledge sources is with individual lenses (or their human filtering). Strategy and goals provide overall guidance for the whole process (Figure 7.3).

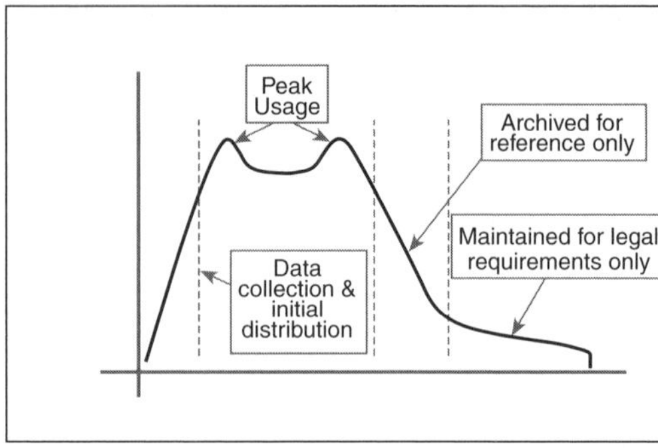

Figure 7.3 Idealized time value of information linear processing cycles.

Information needs be well architected at a few levels. It is this author's beliefs that information needs to be thought of and diagnosed similar to the way in which we think of application code.

Gates (8) stated, "A major shift in the computer industry has made end-to-end business solutions much more feasible. The realignment of the computer industry from vertically integrated vendors to horizontally integrated, customer-driven solutions has brought prices down dramatically and offered more choice." In order to support information horizontally across drug discovery, extended to drug development and further throughout the product life cycle (both within and outside the enterprise), we need to do the same with information and knowledge. Microsoft products are developed according to a blueprint that establishes a single programming model for the future, the Windows Distributed Internet Architecture (DIA). Windows DIA has four parts. The first is a forms approach to the user interface that seamlessly integrates web pages using hypertext markup language (HTML) with more powerful features found in traditional desktop applications. The second is a component object model (COM) designed primarily to manage business logic across a network. The third part provides a universal approach to data storage so that any program can access data in any format and in any location, such as on a hard disk, in a computer database or e-mail folder, or practically anywhere else the data are stored. The fourth is a mechanism that enables computer processing to be done wherever it makes the most sense—on the client, on the server, or some combination of both, or replicated from the server to the client for use by mobile employees.

These same principles and concepts translate to information. In many ways, this can be thought of as a metaconcept applied to the enterprise. Instead of programming rules and standard code, information requires cod-ification and taxonomies. Instead of code objects, let us think about content objects and Web services, as well as content services (Table 7.2).

"Once a company has a blueprint, another architectural imperative is to design programs with a 'three-tier architecture' to separate the logic of a program into three classes: the presentation layer that displays data to the user, the middle layer to encapsulate the business rules of the application—

Table 7.2 Application/Infrastructure vs. Information/Knowledge Parallels

Application/Infrastructure	Information Knowledge Management
Four-tiered approach/ architecture	Four-tiered approach/architecture and framework
Tier 4: Enables computer processing to be done on clients, server, combination of, or mobile	Tier 4: Compartmentalized information on mobile, PDAs, etc.
Tier 3: HTML layer seamless integration to Web pages	Tier 3: Personalized layer through enterprise portal-centric view, taxonomy management, information abstraction
Tier 2: Business rules/ Web services	Tier 2: Information services/enterprise services
Emerging layers Core layers (SOAP, XML, TCP)	Marriage between personal rules and information rules
Tier 1: Universal approach to data storage so any program can access data in any format and location in a database, e-mail, anywhere data are stored, operating system (O/S)	Tier 1: Information stores/libraries and codification Data repository attributes/metadata
Controls—reusable software components add specific functionality quickly (i.e., stock ticker)	Content components define need for code controls (product, people—need for core data schemas); add reusable content quickly
Distributed client–server and topologies	Distributed information—horizontal and extended organization
Coalition of code creators/ owners	Coalition of content creators/owners
Version controls	Version control through work and information flow

(SOAP: simple object access protocol; XML: Extensible Markup Language; TCP: transmission control protocol.)

whether a price cut should apply to an outstanding order, for instance—and the back-end layer that stores and retrieves business data" (8). Gates was, of course, writing about a digital nervous system then and a well-defined framework for how to organize the computer hardware, network, applications, and operating system. How does this apply to information? Well architected IM enterprise services should adhere to an information services framework that allows for an enterprise flow of information. That framework needs to combine a high degree of central control with the flexibility required for information and knowledge management. We no longer need to make the trade-off between these two approaches. Another dimension is that we have information at the personal productivity layer and also the enterprise layer, just as in the old mainframe days, when there was enterprise mainframe and the pendulum shift to PC personal computing.

Over the years, professionals, particularly in the information services area, have seen shifts from the centralized mainframe IT systems to a distributed client–server-based approach. The pendulum shifted from traditional centralized IT controlled systems to decentralized solutions, and now, with the onslaught of so much information via systems, e-mail, eCollaboration, files shares, and so forth, we are trying to find the middle point of the pendulum—a place where owners of information can update and be responsible for their information themselves. However, the management of the information and knowledge is now a critical role for the IS professionals to play in order for there to be some balance for users and owners of the systems to concentrate on their core business processes. In order for this balance to occur, we must take multitiered approaches to the various information levels.

There is a lot to be said about some homogenous solutions for enterprise value. There needs to be a balance of homogeneity and heterogeneity to support central enterprise services and flexibility for innovation and local needs. For example, PC-based servers support thousands of users yet have 90% hardware commonality and 100% software compatibility with a desktop computer. This homogenous platform is one reason PC systems grew rapidly in popularity for servers. Having the same operating system on the desktop and on the server simplifies development and training and establishes a uniform architecture for distributed computing—applications or parts of applications can move from any machine to any other machine. This commonality also makes it easier to connect knowledge workers to existing back-end data systems. Rather than having to have a middleware software component on every one of 10,000 desktop clients, the interoperability layer can run on a few dozen servers that connect the clients to the data tier.

We have seen an evolution in the perception of information systems organizations supporting all industries. This is an important point because this is a call to the IS organizations particularly supporting the drug development processes, as there is a huge need for IS professionals to provide efficient, effective, and value-added services. The creative uses of tools in the knowledge management space are critical for competitive advantage. The bottom line is that the difference does not lie in the tool selected, but in the homogeneity of the tool in support of better integration between broad processes and subprocesses, as well as successful deployment of the tool within a total solution (with information and data management as a base) (Figure 7.4).

Information service framework includes several solutions that are already existent in companies. It is exploitation of these solutions within a program approach that will allow pharmaceutical companies to best utilize the information generated through the drug development process and bridge the content across the research, development, marketing, and sales life cycles.

The recipe and primary ingredient is in focusing on the content. Or, by making use of the evolving technologies, components and web services

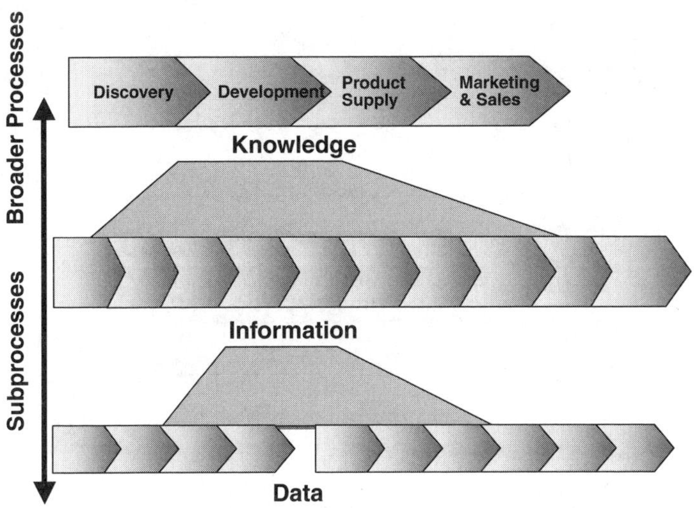

Figure 7.4 Pharmaceutical value chain illustrating where data, information, and knowledge are most prevalent opposite depth of processes.

introduce a new paradigm of IT in managing and maintaining total information services; however, there is promise in the efficiency and effectiveness of integration at the user interface level.

Data and information at the repository level, especially in clinical, tends to be structured data with assigned attributes. These attributes need to be standard. In order for the standards to be fluid, however, there can be costs to maintaining data at the database level. The most important decision-making occurs when understanding the type of information and whether it should be integrated at the database level as opposed to the software/application or user interface level. Most, if not all, of the major pharmaceutical companies have attempted, at varying degrees of success, to integrate information at the database level. It is especially important in discovery and development, where it is so critical to enable information to flow from R&D to submission. So there has been large data warehousing established. There is also large data warehousing established in marketing and sales with customer relationship management (CRM) systems and the like. Data mining can be very effective, but it becomes like spaghetti, especially when information from the discovery side of the house needs to be made available in the marketing side of the house. That is where the benefit of integrating at the application level may be warranted. Often there will be a need for hybrid solutions where information and data are integrated across tiers. Many times information integration is less expensive and less complex (reduces the number of custom interfaces to be built) than integration at the data level (Figure 7.5).

Core to information use and knowledge management at a corporate level are corporate metadata that are a necessity to support modern business organizational functions. As companies are implementing enterprise portals

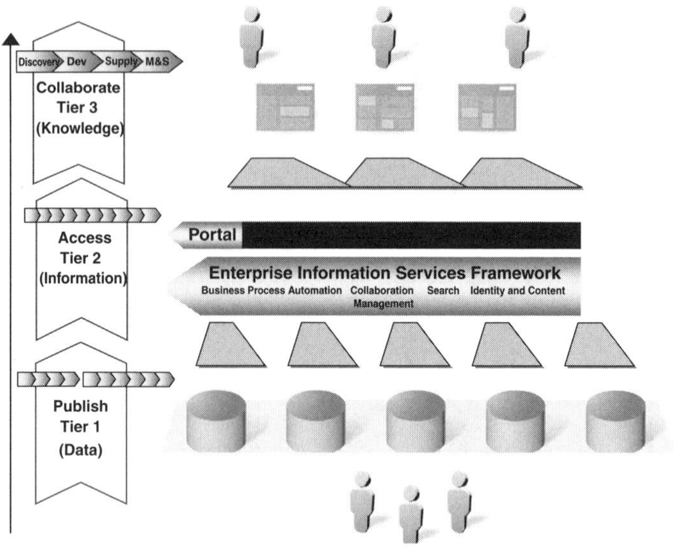

Figure 7.5 Enterprise information services framework enabling depth of processes, breadth across data, information, and knowledge, and varying degrees of vertical and horizontal process integration.

and dealing with heterogeneous repositories while providing a single point of access to enterprise-wide resources and applications, metadata are crucial. Other drivers of smartly standardizing on metadata include the focus in the U.S. on HIPAA (Health Insurance Portability and Accountability Act of 1996) privacy issues and Sarbanes–Oxley (act of 2002, refers to the Public Company Accounting Reform and Investor Protection Act) regulation, and in Europe by comparable compliance requirements. Information within corporations may be structured or unstructured, as well as semistructured, as described by metadata and encoding schemes.

7.5 Knowledge Management Technologies

7.5.1 Overview of Tool Sets

Masses of new data are piling up from the decoding of the human genome among all the other data in support of the pharmaceutical value chain. Pharmaceutical companies are leveraging new technologies ranging from the automated screening of compound potentials to speeding synthesizing and testing of promising chemical compound processes. But to truly exploit the full potential of these technologies, there needs to be a fundamental shift in the way pharmaceutical companies work. CIOs are now being asked to pull back-end systems together with marketing and customer-facing technologies (14). Although the choice of IT tools is of secondary importance to the company's culture and structure as a critical factor enabling knowledge flow, the tools and solutions need to be addressed.

The key to effective tool sets that aid in enabling knowledge management involves a set of tools. Also, the architecture becomes extremely important from not only an infrastructure and application perspective, but also an information architecture perspective. These three lenses must be viewed holistically to best determine how to bridge the architecture on a wide company scale. Some determinations include, depending on the content, when the integration should occur at the data level and when it should occur at the presentation level. In order to extend these tool sets horizontally, companies must get better at the utilization of these internally.

Putting the pieces together within the enterprise information services framework is necessary, together with the enterprise portal, to make up the overall portal framework.

7.5.2 Taxonomy Management

Subject, product, and corporate structures, such as taxonomy and thesauri, are needed, and still emerging, to make the integration and mapping of data and information possible. Navigation through information and knowledge bases requires taxonomy. Taxonomies need to be supported ongoing with the help of taxonomy management tools. Search may be leveraged to generate clustering or suggestions for new or changing categorizations. And taxonomy/taxonomies should permeate across collaborative, directory services, portal, and search categorization browsing.

7.5.3 Collaborative Solutions

When considering collaboration tools, an extremely important consideration is that having more than one collaboration tool within an organization yields out-comes that mimic having more than one e-mail system within an organization.

Real-time collaboration and conferencing vs. project collaboration and document-sharing capabilities.

7.5.4 Directory Services/Expertise Locators

There are a few schools of thought regarding expertise locators. There have been attempts at separate builds of expertise databases. The difficulty here is that no one has the time to go to that extra place just to update his or her expertise information. So these often fail. Then there are the self-service applications, which are a step beyond those expertise databases since there is more incentive for employees to maintain their self-service. There is something in it for them, such as ensuring their paycheck is going to the right place.

Some companies are utilizing an expertise locator through accessing search logs anonymously and linking up employees with employees based on their searches. Other companies are utilizing tools that run across e-mail, again anonymously and securely, to build expertise recommendations for that individual to approve and modify. Since e-mail is used in most companies for everything (at least until the information portfolio is varied and better defined), this seems to be the most promising tool for expertise location. Combining this with a search may perhaps be beneficial. However, one could argue that if someone receives or sends e-mail on certain topics, he or she is more than likely an established expert vs. someone who recently searched on an item. The person that searched is probably searching because he or she wants to learn more.

7.5.5 Search Management

Enterprise search and search components are important elements within the enterprise information services framework. There are advantages in choosing a common search engine across the company. These advantages include the ability to leverage common skill sets, less strain on the network due to a limited number of search crawlers, fewer resources to maintain common catalogs of information, common algorithms, and the ability to categorize information for browsing. There will always be some search engines that are vertical within a solution, particularly at the information and data layer, so many of the enterprise search engines can connect to those various engines and serve as a metasearch across engines. Of course, limiting the number of these helps limit the necessity of resolving the differences between varying result sets. Visualization facilities are available within search tools. These facilities allow heat mapping or visible clustering of hot topics across unstructured information.

7.5.6 Portal Management

There are many types of portals. There are functional portals, vertical portals (vortals), uber portals (portal of portals), and enterprise portals. An enterprise portal provides information at the corporate level (news and announcements, navigation within and outside the organization), as well as access to aggregated information from the divisional and team levels (search and subscription). For any user in the organization, a variety of training, public information, or other material may be available either as a permanent collection of video on demand or as a series of scheduled broadcasts available to any desktop.

At all levels, content may be restricted to specific users or groups, enabling them to assume specific roles (author and reader, for example) over specific sets of content. Given that all servers authenticate from a common directory or set of domains, access control is preserved and respected at all levels of the workspace hierarchy.

7.6 Barriers and Obstacles—Not Just Technology: Organizational Changes

There are principles that need to be adopted, promoted, and enforced within organizations for KM (and its IM precursor) to happen. The top tier of management needs to be interested in order to stand behind the transformation.

Measurements need to be formulated and sustained. Continual learning and feedback are required, as well as ensuring that there is strategic and business impact in areas of focus. Speed the flow of information. "Share information across department lines. Operate as one company, not as separate, competing factions. Your ultimate competitor shouldn't be found across the hall but outside your organization's door" (19).

Maintaining a constancy of purpose while energizing and rewarding employees for sharing information and knowledge is critical to effect change. The processes, governance (especially where there is cross-functional ownership), addressing the sense of reduced autonomy, and controlling and promoting the information ownership and sharing are mandatory. The technology is an important component, but not nearly as important or as difficult, as these change management and cultural issues. The cultural shift is not to be underestimated, and the balances to be acheived with policy, flexibility, coordination, and innovation need to be learned.

It is imperative to develop and implement ways to capture, measure, trade, and appropriate reward.

Culture is often cited as a central issue to KM, as KM is described as a people-dependent activity and largely information (1). People and their interactions create knowledge. Tackling the "knowledge is power" culture requires large changes and strong, consistent leadership over an extended period.

Showing efficiencies and economical value has begun, in most companies, with process metrics and tracking of costs vs. budget regarding websites. Year-to year value analysis of the enterprise's technology assets has also been performed. Commitment from the various areas throughout the company needs to be gained in order to quantify the impact on the vertical business processes. These metrics and more are still developing, especially where metrics may extend across horizontal processes.

7.7 Case Study

7.7.1 Historical Account

SCRIP (17) outlined a historical account of many companies.

7.7.2 Success Factors

Rationalization of Web sites while appropriately aggregating and consolidating websites in an orderly fashion has the potential to show significant savings in overall costs of developing and maintaining sites—not to mention the avoidance of duplicative content or outdated information. IS can take it upon itself to "eat its own dogfood" and reduce the number of sites. Further aggregation can happen across the consolidated sites, and opportunity costs for enterprise services, content components, and enterprise information services framework can prove tangible and intangible efficiencies.

Adopt a hybrid central/decentral approach with flexibility, with the enterprise services central, but the content management, publishing, access, and information life cycles belonging to the business. Enabling the information management life cycle and presentation layer while describing the principles of setting up or utilizing existing information repositories and appropriate metadata, taxonomies are important.

Partnering with a key customer and area of the business to be the first anchor tenant and pragmatically phasing in content, services, events, audiences, and enterprise services make enterprise portal and its framework more tangible.

7.8 Learnings

Management implications were demonstrated in PharmaCorp's experience with KM (3). Some key lessons there included the assurance that knowledge is managed around communities of practice. In other words, it is important not to focus primarily on IT and infrastructure, but also on the process and knowledge interdependencies within the organization. Also, PharmaCorp learned to not focus on explicit knowledge only, but to question each piece of information with a business mind-set. It is important to ask what would happen without this information. Leading from the top is key while making sure appropriate rewards and reward processes are in place.

7.9 Future Look

All the hype about knowledge management and variances on the definitions, applications, and so forth seems to indicate infancy. The infancy began with an overall vision and recognition that companies need to do things differently in a knowledge era and even more demanding business environments. There are still proponents and opponents of knowledge management, but if we step back and look at the time spectrum, there has been much accomplished. Several examples of vertical KM have been illustrated across the pharmaceutical value chain. These examples suggest that KM can be successful. We all know that huge shifts in the way things are done take time. For those who expect immediate results, there will still be an anxiety and a "prove it to me" attitude—especially for the naysayers of KM. But now that companies have some good vertical KM solutions, there is a foundation to build on and enable more horizontal KM. As good examples in that arena evolve, the true value of KM will be more apparent, perhaps even so apparent that it will be considered business as usual and an infrastructure of sorts itself.

To move toward a component-based infrastructure and a more horizontal business climate (within R&D and other areas of the pharmaceutical value chain, as well as extended throughout with partners and alliances), we need to treat information on an enterprise component-based scale to best utilize the technology and, most importantly, deal with the new business environment. Compartmentalizing information into smaller, more granular units can provide competitive advantage for reuse and maintenance. Pharmaceutical companies are dependent on developed science-based processes where the locus of critical knowledge conversion occurs upstream. Downstream processes are dependent on knowledge codification that occurs in the upstream process.

Managing content services along with application and infrastructure services is best through an overall portal framework, which is inclusive of enterprise information services. Business processes determine what should be a distributed or central commodity or proprietary solution. Value-added measures here include opportunity costs and total costs savings. In order to fuel economic growth, understanding how to cultivate and manage knowledge and information resource requirements becomes increasingly important. This leads to better, more integrated business processes and sub-processes. Companies must better manage the visibility into portfolio of drugs, increase efficiencies with cross-geographic and cross-company teams, and improve KM and access to the most current data and information. Measures need to be demonstrated through streamlined product life cycles across the pharmaceutical value chain, reduced cost of research and development, and reduced number of inefficient compounds early in the value chain. One may still argue the value since measurement is in its infancy, but companies need to continue improving, and the risk of trying KM is a gamble this author feels is worth the bet.

Glossary of Terms

Collaboration—The act of networking through people to solicit, integrate, or meet to share or exchange ideas. eCollaboration is simply networking with people and communities through virtual teaming online.

Content—Generically, any information that is available for access to employees within the company. Content can exist in the form of websites, applications, e-mail, databases, etc.

Content services—Content that is a high candidate to make available once and use many times.

Data—Sets of discrete, objective facts about events. Knowledge that is explicitly captured.

Enterprise information services—Information management, applications, and technology (end-to-end solutions) that are offered as a central service within the corporation.

Enterprise portals—A means of presenting a business professional with clear, consistent, and ordered access to multiple information sources and applications via a corporate web.

Events—Content owned by various functions within the company integrated through presentation to perform an action or process.

Heuristics—Rules of thumb or educated guesses learned from experience that experts use in solving problems in their domain of specialty. An independent discovery method for learning.

Information—Evaluated data for specific individuals working on a particular problem at a specific time and for achieving a specific goal. Also, selected data for reducing the amount of ignorance or the range of uncertainty in the mind of the decision maker. The selection of a given message from a set of possible messages that informs the recipient of potential value. Data that make a difference and knowledge that is explicitly captured.

Informatics—Information management applied to specific medical and scientific data.

Information management—The ability to access, share, and exploit content within the enterprise while the content itself is supported by an information life cycle.

Innovation—The successful exploitation of ideas to create a new, useful offering of product or service.

Knowledge—That within and between the minds of individuals. Tacitly possessed and has the capability to add value.

Knowledge base—That body of facts, rules, principles, laws, and heuristics that forms the basis of a knowledge system.

Knowledge-based system—A decision support system containing a knowledge base and an inference engine.

Knowledge engineering—The acquisition, organizing, and designing of a knowledge base consisting of facts, rules, laws, principles, and heuristics.

Organizational effectiveness—The degree to which the organization achieves its goals or objectives. Also, the ability of the organization to exploit its environment in the acquisition of scarce and valued resources.

Portal—Can range from a website's home page to an enterprise portal. An enterprise portal is one window where all available content, applications, processes, and collaborative resources within a company can be viewed.

Search—The ability to search across a company's content assets regardless of content store, media, or reach. Also, the ability to find content within a specific information store.

References

1. FMR Armbrecht, RB Chapas, CC Chappelow, GF Farris. Knowledge management in research and development. *Res Technol Manage V* 44: 2001.
2. Anonymous. *KM and the Pharmaceutical Industry: A Rountable Discussion—Part 1*, Vol. 12, No. 6, June 2003.
3. Anonymous. Wella, Unilever and "PharmaCorp" Have Mixed Experiences with KM. *Strategic Direction* 19: 2003.

4. LB Cardinal, TM Alessandri, SF Turner. *J Knowledge Manage* 5: 2001.
5. S Davis, C Meyer. *Blur: The Speed of Change in the Connected Economy.* MA: Ernst & Young Center for Business Innovation, 1998.
6. C Downey, D Greenberg, V Kapur. Reorienting R&D for a Horizontal Future. 46: 2003.
7. P Drucker. *Tasks, Responsibilities, Practices.* New York: Harper & Row, 1974.
8. B Gates. *Business @ the Speed of Thought.* New York: Warner Books, 1999.
9. W Hamilton. e-Ticket to global harmonization. *Pharm Executive* 22: 2002.
10. B Lev. Knowledge management: fad or need? *Res Technol Manage* 43: 2000.
11. R Marcus, B Watters. *Collective Knowledge.* Redmond, WA: Microsoft Press, 2002.
12. H Mitsuhashi. Uncertainty in selecting alliance partners: the three reduction mechanisms and alliance formation processes. *Int J Organ Anal* 10: 2002.
13. I Nonaka, H Takeuchi. *The Knowledge-Creating Company.* New York: Oxford University Press, 1995.
14. S Overby. They Want a New Drug: In the Race to Develop New Pharmaceuticals More Quickly, Companies Are Introducing New IT Tools, but the Tools Can't Do It Alone. CIOs Need to Change the Way People Do Their Work. *CIO*, October 15, 2002.
15. P Schoderbek, C Schoderbek, A Kefalas. *Management Systems: Conceptual Considerations.* MA: Irwin, 1990.
16. T Studt. Knowledge management is key to improving drug R&D cycles. *R&D* 45: 2003.
17. S Ward, A Abell, with TFPL LTD. *Scrip Reports: Mobilising Knowledge: The Pharmaceutical Industry Approach.* London: PJB Publications, 2001.
18. D Wesley. Retaining workers in the knowledge economy. *Inf Outlook* 35–40, 2000.
19. C Wick, LS Leon. *The Learning Edge.* New York: McGraw-Hill, 1993.

chapter 8

Understanding the Value of Research

Paul Bussey, Jo Pisani, and Yann Bonduelle

Contents

It is well known that to grow, life science companies must continuously invest in research with the goal that these investments will help to discover new therapies for unmet medical needs. In recent years, significant advances in medical science have been achieved, such that there is an increasing wealth of investment choices available. The challenge is to select those investments that will produce the best return.

Investment opportunities include drug discovery programs, discovery technologies, drug platforms, informatics systems, and research infrastructure. Of course, more often than not there are far more investment opportunities than money available. Faced with limited resources, management must rationalize and make choices between alternative investments. To help make these choices, most investors across industries tend to measure the value, expected return, and risk of each investment opportunity. Based on these measures, investors can make informed decisions as to which investments they prefer given their different objectives.

There is often resistance to putting financial valuations against early-stage R&D projects. Two objections are most commonly expressed. First, the R&D process, by its very nature, is based on serendipity in both the likelihood of scientific success and commercial potential, and it is therefore difficult to quantify what some would argue is a highly random process. Second, any attempt to quantify financial returns on R&D at such an early stage will only stifle creativity. These are valid concerns when the valuation approach is applied too rigidly. However, if a pragmatic approach is adopted, early valuation will assist with decision making as the project progresses. Our recommendation is that valuation be used to establish a baseline. An appropriate amount of time should be spent to understand the value that may be generated by the project, the cost it consumes, and the risk it bears. This baseline can then be used to assess the impact of further developments or options that emerge during the discovery process.

The discussion above applies to the "old world" of drug discovery: the scattergun approach where a collection of promising targets is investigated in an *ad hoc* fashion. We are now in the era of rational drug design, which should result in more focused research with more predictable results. In this new world, early valuation helps focus on fewer options earlier. In this new world, it is even more important to understand which are the higher- value options before placing bets in one area, and this requires ensuring that the baseline is in place and is credible.

From our work in this area, we observe that valuation approaches are well developed in the late development function (namely, clinical trials). The rationale for this is clear: investments in clinical trials are substantial and regulatory processes force drugs to be launched in specific well-identified indications. It is then possible to evaluate potential market size and market share with a relatively high degree of robustness.

Most, if not all, leading R&D organizations have used valuations in late-stage development for many years. Some of those organizations are now applying valuation approaches earlier in the process to drugs in preclinical development. Very few, however, measure the value of projects in research.

This situation will be changing, and this chapter, we hope, will help convince researchers and managers that systematically spending a little time evaluating investments with appropriate quantitative valuation approaches brings three key benefits:

- It produces a baseline reference value for each investment, allowing researchers and managers to understand the value, cost, and risk trade-offs associated with that investment.
- It allows competing investments to be compared, thanks to a consistent evaluation framework.
- It helps the organization remember why some investments have been made or turned down. For those made, it provides a simple starting point of the evaluation dossier that will follow the investments throughout its life.

These benefits contribute to making research more productive by helping direct cash and people to the right investment. This is why, in our practice, despite natural organizational inertia, we relentlessly promote the creation of an appropriate (light, credible) valuation structure and processes to get systematic quantitative valuation of investments in research.

In this chapter we discuss the reasons for conducting valuations in research, outline the types of investments that are made in drug discovery, discuss some of the challenges in implementing valuation, and detail valuation approaches. For readers interested purely in valuation approaches, we recommend that you proceed directly to Section 8.3.

8.1 Why Do We Need Valuations in Research?

Management's responsibility is to manage effectively a business's assets or investments to maximize their return for the benefit of shareholders. Deciding which investments to pursue and the strategy for exploiting them requires a method for assessing and comparing the available investment options. Valuations provide a basis for measuring the relative potential of alternative investment options and therefore build the foundations for making investment decisions. Producing robust, reliable, and insightful valuations can help a company make better investment decisions, leading to an enhanced business performance.

8.1.1 Making Investment Decisions

At their simplest, valuations indicate whether an investment is worth pursuing. If, taking into account the amount that needs to be invested, the valuation of an investment opportunity has a positive value, then it will generate returns that meet or exceed an investor's requirements, implying that the investment should be pursued. However, if the value of an investment opportunity is negative, then it will not deliver the investor's required level of return and should be ignored. In this respect, investments in drug research are no different from any other type of investment. Life science companies must carefully consider how much they will get back in the future for their research investment today and what is the likelihood of achieving that return (their cost of capital) before committing their resources.

Consider a pharmaceutical company that is deciding whether to invest in a new drug research facility. A valuation analysis would help the company to understand the potential benefit of the investment by providing a measure of the expected financial reward. It would require the company to identify the potential investment outcomes, assess the key factors driving those outcomes, and determine their likelihood and the associated value in each case. This should lead to a considerably enhanced understanding of the opportunity and enable the company to make a better, more confident investment decision.

Any investment decision needs to be made against the background of a company's investment objectives, including investment time horizon, risk tolerance, financial constraints, return objectives, and business strategy. There are likely to be significant differences in the outlook and attitude toward investment between different companies, particularly between big pharma and small biotech. With a large income, strong bank balance and secure position, big pharma can afford to take a long-term view of its investment activities. This enables it to make higher-risk investments with longer payback times that in the long run may produce greater rewards than shorter-term investments.

In contrast, a cash-strapped biotech that is rapidly burning through its remaining capital may not be able to afford long-term or high-risk investments. This may force a biotech to make investment decisions that limit the potential reward. For instance, rather than discovering and developing drugs using a proprietary drug platform, a small biotech may have to license out the technology to other companies and receive milestones and royalties in return. This is likely to produce significantly less reward than going it alone.

The investment preferences of the company can be captured in the valuation analysis through the explicit incorporation of the probability of success and setting the rate at which returns are discounted to reflect the time value of money.

The discount rate or cost of capital is a critical risk factor in any valuation analysis. It represents the required level of return an investor expects from an investment, given the perceived level of investment risk, in order to justify making the investment. The higher the discount rate, the lower the valuation. The impact of discount rates is most pronounced for investments with long payback times where distant cash flows are more heavily discounted. This produces an increasing bias toward investments with shorter paybacks as the discount rate increases. Significantly, the discount rates for small biotechs tend to be larger than those for big pharma, resulting in different value perspectives when evaluating potential investment opportunities.

8.1.2 Portfolio Management

Making investment decisions on a single investment opportunity can be relatively straight forward, assuming you can establish a robust valuation. You either invest in the opportunity or you do not. However, investment decisions become significantly more challenging when you are faced with a number of alternative investment options, and your investment choices are limited by your resource constraints. These circumstances require a process for making the best overall set of investment decisions. Portfolio management is the process of allocating investment to best implement your corporate objectives given your available resources.

Effective portfolio management assists the implementation of corporate strategy by identifying, developing, and managing the best set of investments available. If performed well, portfolio management increases a corporation's likelihood of successfully delivering its strategy and meeting shareholder expectations. Conversely, poor portfolio management will lead to failure to deliver shareholder expectations.

Portfolio management is most commonly applied to the selection of project investments in research and development portfolios. Life science companies often have more projects in their R&D portfolio than they can actually afford to fund, forcing them to choose between the projects they will fund and those they will stop or delay.

Though many different factors need to be taken into account when performing portfolio management, such as strategic fit and competitive positioning, the expected financial performance of each investment is usually the critical factor. Valuations, together with the various investment performance measurements based on them, provide the cornerstones for portfolio management. They allow the comparison, ranking, and prioritization of investments against specified constraints to meet a company's financial objectives. This provides management with the information and guidance to make the best investment decisions across its portfolio of investment opportunities.

Traditionally, if you asked a pharmaceutical manager about portfolio management, he would be likely to assume that you were talking about drug development. However, pharmaceutical companies also make substantial investments across their business, including sales and marketing (S&M), supply chain, informatics, mergers and acquisitions (M&A), licensing, and research. Though most large pharmaceutical companies perform portfolio management within development, at least within late-stage development, it is far less common for portfolio management to be applied within research or other activity areas. It is even less common to find portfolio management performed across functions.

Figure 8.1 shows, for the main functional areas in pharmaceutical companies (research, development, central sales and marketing, manufacturing, informatics, licensing, and local sales and marketing), the value that is

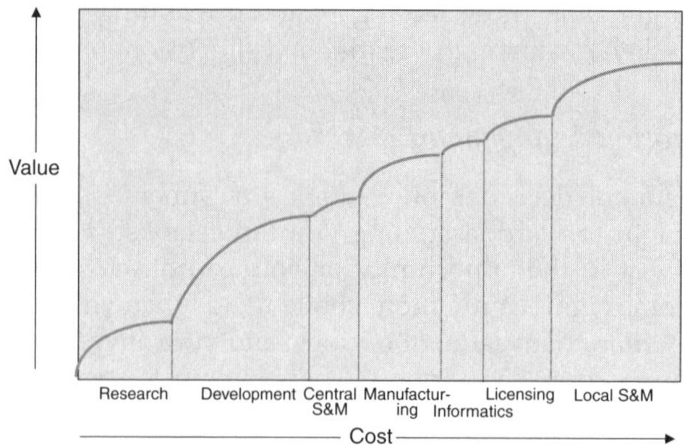

Figure 8.1 Portfolio thinking supports enterprise-wide resource allocation discussions and decisions.

expected to be delivered by each area given a certain level of investment. This curve is just an accumulation of the efficient frontiers for each of the activity areas. An area's resources are the width of its slice on the x-axis; its value is the rise of the curve over that interval.

Senior management's responsibility is to invest resources across and between different areas to support the implementation of corporate strategy. This requires the development and application of portfolio management across the organization, including research, and not just within development.

8.1.3 External Collaborations

In the search for new drugs, many companies are increasingly looking beyond the boundaries of their own organization as the productivity of their internal research has fallen. This has fueled a proliferation in the number of research collaborations through licensing deals, joint ventures, equity investments, and acquisitions. The precise terms and conditions of most deals are agreed on through negotiation, but the question for both parties is whether they are getting a good deal. Deals can involve complex financial arrangements with combinations of up-front and milestone payments, laddered royalties, loans, and equity investment. Determining and agreeing on the best deal structure for both parties can be challenging.

Valuations enable a comparison of the financial performance of alternative deal structures from the perspective of both the licensee and licensor. The valuation of a deal typically starts with the analysis of an investment's potential as if it were already part of the acquirer's portfolio. Based on the expected risk and return of the investment, the value of a deal under alternative deal structures and parameters can be determined (Figure 8.2). This provides a basis for establishing the parameters for deal negotiations.

Again, deal valuations need to be considered in light of the objectives and constraints of the parties involved. Consider the case of a licensing deal for a

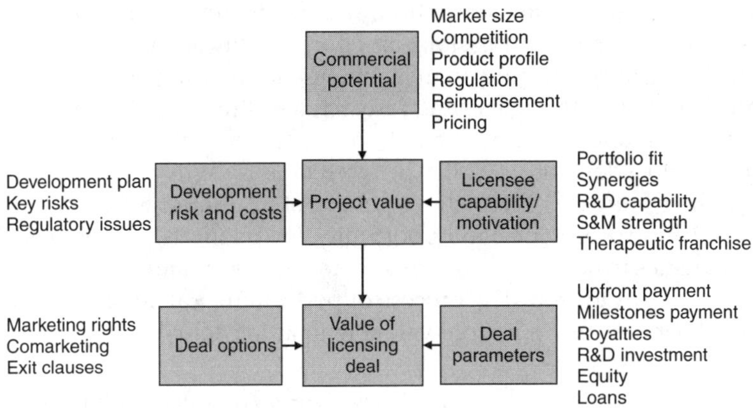

Figure 8.2 Evaluating licensing deals.

group of drug targets. One potential deal structure may be based on relatively high royalties if a drug based on the target reaches the market, but may promise only moderate up-front and milestone payments during R&D. This is likely to have a greater value than an alternative deal structure with more significant up-front and milestone payments for progress achieved through R&D, but smaller royalties. If you are a cash-strapped biotech with a need for capital in the short term to keep your business going, you may prefer the lower-value deal with greater up-front payments and forego the potential of more lucrative royalties. The question is: What is the best deal for your business? Valuations help you to answer this question by evaluating the implications of alternative deals.

8.1.4 *Increasing the Value of an Investment*

Though the principal question for management is whether to make the investment, many other difficult questions need to be addressed. Management must also decide how much to invest, when to make the investment, and how best to manage the investment to meet its objectives. To make these decisions, management must answer a number of key questions:

- What is it worth? What are the risks/uncertainties? What are the strategic options? Is it a good deal?

Valuing a research investment is about not only obtaining a measure of its comparative worth, but also providing a better understanding of the investment itself. Developing a better understanding of a potential investment and the source of its value enables management to manage the investment better and so increase its worth. For instance, it may be possible to develop risk mitigation strategies that increase the probability of success for a compound or devise alternative research and development plans that will lead to a more valuable product profile.

Unfortunately, it is neither easy nor always possible to identify and evaluate a broad range of potential options for an investment, the possible outcomes associated with each option, and the likelihood of achieving them. But to the extent that it is possible, an effort should be made to understand both the key factors affecting an investment and their impact upon its value.

The process of valuing existing research investments provides a check on the justification and viability of research initiatives. It forces management to ask critical questions and challenge expectations. Why are we doing this research (investment)? What will we get out of it? Should we be doing something else? It also provides an opportunity for previously undisclosed concerns and issues to be raised by stakeholders. This can justifiably close a project that no one dares to stop. The process of performing valuations can produce additional benefits that are more valuable than the actual valuation itself.

8.2 What Type of Drug Discovery Investments Do We Make?

Pharmaceutical companies are faced with research investment opportunities ranging from drug platform technologies, through informatics systems and drug discovery tools, to actual discovery projects. These investments can be made through internal investment, licensing deals, and acquisitions. Different types of research investments have different valuation considerations that need to be incorporated into a valuation analysis.

8.2.1 Drug Discovery Projects

In attempting to discover new therapies for a disease, researchers must navigate a vast and complex maze of different possible disease pathways, which may be linked with many different therapeutic targets susceptible to different forms of therapy. This creates a huge opportunity for the discovery of new therapies. However, finding therapies within this maze is very challenging and knowing where to invest your research effort for the best return is difficult. Valuations provide management with a tool for assessing what are the potential benefits, risks, and rewards of different research initiatives and can help them decide where best to invest their money.

The industrialization of research has resulted in organized research programs, focused on developing specific therapeutic techniques aimed at specific therapeutic targets. This sets a clear objective for the research and provides a basis for performing a valuation analysis. Any valuation analysis of a research program will need to take into consideration the potential value of disease areas being tackled, together with the strengths and weaknesses of the therapeutic approach. Evaluating the strengths and weaknesses of a research program means comparing existing and future therapies for the disease area:

- What advantages will this research bring? What value will be placed on the expected therapeutic improvements? What is the downside? How will this be viewed by the market? What impact does it have on the value?

Answering these questions leads to an understanding of the potential value of a research program.

At the same time, the potential risks involved in achieving the objective of a research program must be considered:

- What are the technical risks associated with this type of therapeutic approach? What is the likelihood of this therapeutic approach working in this disease area? What are the commercial risks? Do we have the experience and capability to be successful?

Combining the evaluation of risk with an assessment of potential reward produces a valuation, providing management with insight into the potential of a research program. The valuation produced not only provides management with the information necessary to make informed investment decisions, but also provides a basis for changing investment plans to increase their potential value.

8.2.2 *Discovery Tools*

The industry's desire for greater research productivity has fueled the creation of drug discovery factories costing hundreds of millions of dollars and offering a diverse range of technologies from high-throughput screening (HTS) systems to bioinformatics tools. Choosing the optimal technology investments to meet a company's drug discovery objectives requires careful consideration of the value created by each technology.

The considerations needed to value technology investments differ from those used to value drug discovery projects. The inherent value of technology investments is not necessarily in the technology itself, but in the impact that it has upon the drug discovery process. The focus is on how technology investments increase the overall value of the drug discovery operation, rather than the actual value of individual drug discovery projects. New technology investments can improve the drug discovery process through increased throughput of drug compounds, reducing the drug discovery processing time, increasing the compound hit rate, and improving the quality of lead indications.

Moving from HTS-based technology to ultra-HTS should lead to an increase in the number of compounds screened against targets with the potential to produce a larger number of target hits. This investment increases the productivity of research and the number of new compounds entering development. Assuming that the quality of additional target hits is consistent with previous results, the value created by investing in the technology is based on the value of the additional compound hits. Investing in new technologies that increase the hit rate should create value in a similar manner. The valuation of such technologies is focused on how many more drug development candidates the investment would produce than the status quo.

Improving the quality of drug development candidates produced can also significantly increase the value of drug discovery operation. Valuing technologies that produce such improvements requires careful consideration of a number of factors. First, improving the quality of development candidates can lead to a reduction in failure rates during development. This both reduces the cost of producing a new drug and increases the number of new drugs that are successfully developed, resulting in a double value benefit. Second, improving the quality of development candidates can produce better product profiles, leading to higher-value drugs. When valuing technologies that improve the quality of drug candidates, attention should focus on the expected improvement in the commercial potential of candidate drugs and the development success rates.

Technologies that shorten the drug discovery process can significantly increase the value of drug development candidates by reducing the time to market and increasing the period of market exclusivity. This brings investment returns closer and increases their size. Reducing drug discovery times can also increase drug candidate throughput, resulting in value increases through greater productivity.

In reality, technology investments are likely to produce a combination of performance improvements that create value in different ways. Though different technologies can increase the value of the drug discovery operation by different means, the underlying basis of any valuation is the value impact on the drug candidates produced. Therefore, the valuation of drugs within discovery underpins the valuation of drug discovery technologies.

8.2.3 Drug Platforms

Advances in biotechnology in recent years have led to the creation of many alternative approaches to drug creation. In addition to the traditional methods of combinatorial chemistry, new approaches adopted include gene therapy, antibodies, and antisense therapies. In the process, many different enabling technologies have been developed for creating these new therapies, predominantly driven by start-up companies, such as Abgenix and Antisense, that have grown into major drug platform suppliers. At the same time, dedicated target identification and compound discovery companies have created libraries of potential drug targets and compounds. Pharmaceutical and biotech companies focused on discovering new therapies are presented with a wide array of alternative platforms for creating new drugs. As with other types of research investments, the challenge for research-based companies is to select the right drug platforms to meet their discovery objectives and resource limitations.

Evaluating investments in drug platforms combines the considerations used to evaluate both drug discovery programs and drug discovery technologies. An initial evaluation may consider which disease areas and types of targets will be susceptible to drugs developed using the platform. The next stage is to understand the value associated with these disease areas and the potential rewards arising from new therapies in these areas and assess the potential advantages and benefits of drugs developed using the platform over therapies currently on the market and others in development. What

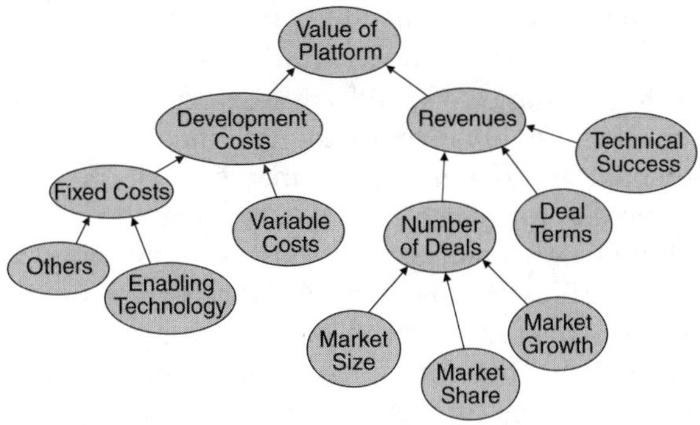

Figure 8.3 Platform value from vendor perspective.

potential efficacy and safety benefits will they produce? This must be balanced with an assessment of the risks involved in developing new drugs using the platform and the chances of success.

The productivity and efficiency of the approach taken to discover new drug candidates must also be assessed. Critically, how long will it take to discover new drug candidates and how many are likely to be produced? Overall, the objective of the evaluation is to determine what value a drug platform will create not only within the discovery operation, but for the company as a whole.

Evidently, in making deals with drug discovery companies the benefits of their technology over alternative and competing technologies are critical. The valuation perspective is slightly different for platform technology vendors (Figure 8.3). The value of a platform is driven by how many deals can be negotiated and on what terms. A platform technology vendor's remuneration may be based on how many drug candidates are produced using a platform, rather than the value (quality and application) of those candidates.

8.2.4 Challenges in Valuing Research Investments

Historically, pharmaceutical and biotech companies have not valued their research investments or managed them using a valuation and portfolio analysis approach. There are a number of reasons for this. Often, there is a lack of organizational buy-in to the benefit of performing valuations within research. Research valuations are perceived as unreliable, time consuming, and not providing a meaningful measure of value. This attitude is largely driven by a belief that it is not possible to produce robust valuations that can be relied upon to make investment decisions. The difficulty in valuing research investments is driven by two key uncertainties: the potential outcomes and the probability of achieving them. This can result in very wide estimates of value that are of little help in making investment decisions.

8.2.4.1 Uncertainty around the Potential Outcome

The outcome of research investments can be difficult to predict, especially for novel, untested therapies. Consider a research program focused on the creation of monoclonal antibodies for the treatment of cancers. Though the commercial potential of different cancers may be well understood, it may be very difficult to predict for which cancer indications the research will be successful. This immediately creates significant uncertainty as to the potential value of the research program. Furthermore, the benefits and disadvantages of the antibodies discovered over competing and alternative therapies may also be hard to establish. Therefore, not only is it difficult to ascertain in which indications a research program may be successful, but also how successful it may be within those indications. This creates challenges as to how the uncertainty and variability in the outcome within a valuation can be captured.

8.2.4.2 Predicting the Probability of Research Success

The main driver of value in life science investments at all stages of research and development is risk. The probability of successfully developing a drug from target identification to market launch is typically less than 1%. Drugs can fail at any phase of development, for any reason. While low probabilities of success in researching and developing new drugs are accepted as a fact of life, some research investments are riskier than others. A measure of the level of risk associated with each investment is required such that this can be weighed against the potential value. The challenge is to understand the potential risks of a research investment and from that formulate an opinion of its likely success.

Estimating the probability of a drug successfully reaching the market requires careful analysis of the drug's risk attributes:

- Why is a drug likely to have a greater or lesser safety risk? What do we think the pharmacodynamic and pharmacokinetic properties will be? What indicates that the drug will have significant efficacy?

Answering these questions enables the development of a risk profile for the drug, from which an estimate of the probability of success can be determined. This is the approach commonly taken during drug development where significant knowledge and experience have been gained about a drug.

However, for drug candidates that have not yet been discovered, it is not easy to determine a risk profile and likelihood of success. Given limited evidence to formulate an opinion, experts may not be able to forecast the potential attributes of drug candidates. This makes formulating a reasonably reliable view of the probability of success for drug candidates difficult. Similarly, determining the productivity of a research program, whether it is the number of drug candidates produced or targets identified, can also be challenging.

The uncertainty around how many drug candidates may be discovered, the probability of them reaching the market, and the possible value of those drugs combine to produce significant variability in the valuations of drug programs.

8.2.4.3 Resistance to Valuing Research Investments

Valuation analysis is not natural territory for most people working in drug discovery. Researchers are generally not as commercially focused or attuned as their cousins in drug development and are much happier working in the lab. Questioning and challenging the purpose of research on a commercial basis is not what researchers are used to and probably frightens them. This produces resistance to implementing a valuation approach for assessing research investments. Given the technical difficulties, it is all too easy for people to criticize valuations. Therefore, achieving organizational buy-in is essential in order to successfully implement a valuation process for research investment.

8.3 Valuation Approaches

In discussing valuation approaches, our focus will be on valuing drug discovery projects rather than investments in technology. The valuation of drug discovery projects underlies the valuation of all other major investments that are made within drug discovery.

8.3.1 Approaches for Valuing Drugs in Development

Most pharmaceutical and biotech companies perform valuation analysis upon drug development projects from first time in human (FTIH) onward. Some companies have implemented processes for valuing drug development investments from the lead selection stage onward. The valuations of drug development investments or projects are predominantly performed to support portfolio reviews of drug development investments and, in the context of the annual budgeting process, provide one of the key inputs into the decision on whether to continue, stop, or postpone drug development projects. However, valuations are also commonly performed when considering potential licensing deals for development drugs, as well as acquisitions or disposals of drug development businesses.

Understanding how development investments are evaluated provides a strong basis for evaluating research investments as well. All of the considerations for evaluating development projects are also necessary when evaluating research investments. Development is simply a later stage in reaching the same successful investment outcome as a research project, i.e., the development of a marketable drug and similar considerations apply, for example, target market, probability of success, and commercial advantages.

The most common approach to valuing drug development investments is to determine the expected net present value (eNPV) of the investment using a decision tree approach. Decision trees (Figure 8.4) provide a convenient approach for valuing investments around which there is significant uncertainty regarding the future outcome and associated cash flows. They allow the value of different outcome scenarios to be explicitly captured and then combined, using probabilities for the different outcomes occurring, to

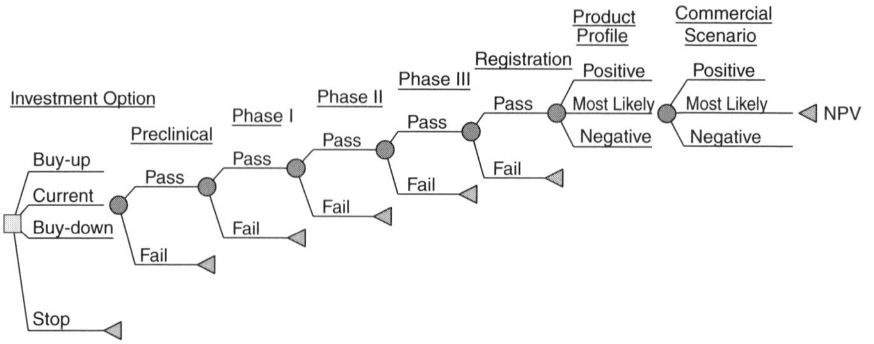

Figure 8.4 Decision tree for valuation of drug development project.

produce an overall average or expected value for an investment. This makes the decision tree approach highly suited to the evaluation of drug development projects where projects can potentially fail at any phase of development and may lead to drugs of different quality competing in a range of different market environments.

Most development investments, particularly drug development projects, have a specific end use or application. Drug development projects are usually focused on the development of a drug for one or more specific indications, for example, the development of a drug for chronic obstructive pulmonary disease (COPD). In these circumstances, the starting point for valuing a drug development project is to understand the disease indications that the drug targets. The objective is to determine the potential value of each indication and provide a basis for determining what share of that value the drug under development may be able to capture.

The epidemiology of each disease indication should be assessed to provide a forecast of expected disease prevalence and incidence over the period the drug being developed is expected to be marketed. For the purposes of a valuation, epidemiology forecasts are usually produced at a regional or country level, with a focus on the highest-value markets (i.e., U.S., Europe, and Japan). Where possible, the disease population should be segmented into definable groups that have different therapeutic needs, such as by disease severity or stage of development. Segmenting the disease population is important since treatment algorithms are usually developed by physicians that recommend specific drugs for treating specific disease segments depending on the perceived needs of patients within each segment and the alternative therapies available for treating patients. The unmet needs of patients within each disease area should also be analyzed, along with the value placed upon them. Measuring the value of different patient needs may involve an economic cost–benefit analysis of satisfying those patient needs, together with an assessment of trends in the pricing and use of existing therapies. The objective is to produce a foundation for estimating the price the market will be willing to pay for the new drug and the potential market uptake, given the drug's potential for satisfying different patient needs.

Having determined the potential value, segmentation, unmet needs, and value drivers within the target market, it is important to assess the competitive environment. The level of competitive threat posed by rival therapies for a market can have a significant effect on a drug's commercial success. Developing an understanding of the potential commercial threats a drug may face in the future is a key step in estimating its potential sales.

Though the strengths and weaknesses of existing therapies within a specific disease area may be well understood, the likelihood of new therapies currently in development reaching the market and their potential impact may be highly uncertain. This results in a wide range of possible competitive environments for our drug, which cannot all be feasibly evaluated. The common approach to evaluating such a wide range is to select a few representative scenarios that cover the majority of likely competitive environments. Typically, a most likely or expected competitive environment is determined, together with an upside and downside scenario. Competitive environments are often defined in terms of the number and classes of competing drugs on the market over the period that a drug is to be marketed. A product profile is developed for each class of competing drug. A product profile should define the performance of a drug class by key attributes such as efficacy, safety, formulation, administration, and so on. The product profiles defined for each competing drug class provide a background for estimating the market share of our drug in each competitive scenario.

When developing a drug for a specific indication, most pharmaceutical companies define a target product profile for the drug. However, assuming development success, it is highly likely that the actual profile of the final drug will differ from the target product profile. Potentially, a range of different product profiles for a drug could be achieved at the end of its development, each with a different commercial potential. The profile of a drug is one of the key factors in determining its market uptake, together with the level of competition and the unmet needs within the market. Capturing the uncertainty around the potential product profile of a development drug and its impact upon the drug's commercial success is an essential component of any valuation analysis.

It is not possible to evaluate all of the different product profiles that could be achieved through the development of a drug. Here too, representative product profiles should be selected so that the differentiation between their attributes creates a significant difference between their commercial potentials.

Drug profiles should be defined in terms of the key attributes that will drive a drug's market success, including dosage, efficacy, safety, regimen, and method of delivery. Pricing can also be specified as part of the drug profile, but may vary according to the drug's competitive environment. Some approaches specify performance of individual attributes relative to the current or expected gold standard therapy at the time of market launch.

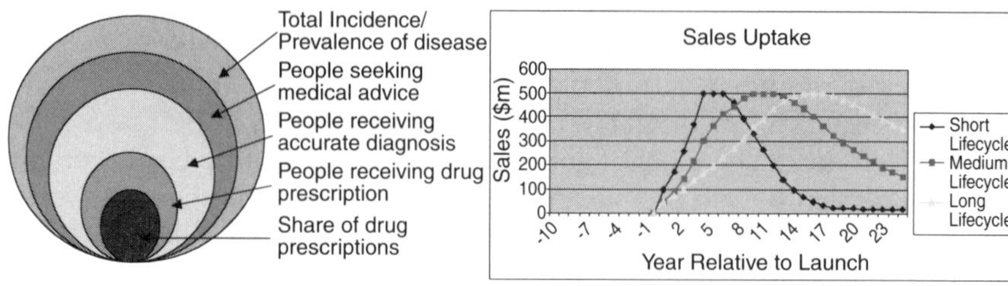

Figure 8.5 Forecasting sales.

The same approach should be used to specify the performance characteristics of competing classes of drugs when defining future competitive environments. This enables a direct comparison of the relative strengths and weaknesses of our drug against its competitors, supporting the estimate of the drug's sales in different competitive environments. Forecasts of drug sales can be formulated in different ways, but commonly are determined as a function of the expected market share a drug will attain, together with the forecasted market size and drug price (Figure 8.5).

An expected average market forecast for a drug is obtained by combining the probability-weighted forecasts for different product profiles and competitive environment scenarios.

When developing a drug there may be several alternatives and distinct development plans to follow, each leading to different product profiles, possibly for different indications. Though often there will be many changes to the individual trials within a development, the target product profile will largely remain the same. However, there may be an opportunity to take an alternative development approach that will lead to a different target drug profile with greater commercial potential. In evaluating a drug in development a company may wish to determine which development approach it should adopt in order to create the most value.

The development plan for a drug provides the basis for assessing two key elements in a drug's valuation: the development costs and the probability of development success. Development costs are normally specified by major development phase, e.g., preclinical, phase I, phase II, and so on. All costs associated with a drug's development are usually considered, including internal costs, external costs, and milestones. The probability of development success is also usually evaluated at each major development. This approach maps probabilities onto the key development outcomes that determine if and when to stop a drug's development, more accurately reflecting the potential outcomes for a development investment.

The probability of development success is the key driver of value for a drug during its development. A drug entering phase I may only have a one in ten chance of reaching the market, making it highly likely that the drug will fail during its development after having consumed significant costs that produce a negative value. Determining the probability of success for a

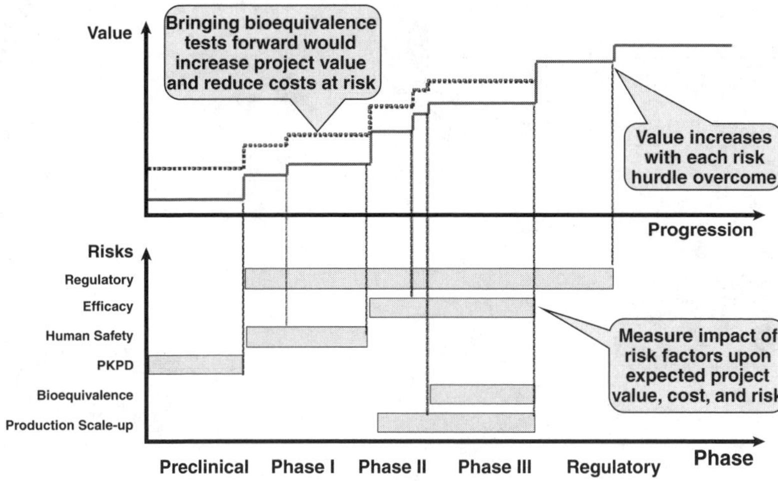

Figure 8.6 Risk mapping diagram. (PKPD: Pharmacokinetics/Pharmacodynamics.)

development project requires understanding the key risk factors at each stage of development. Based on the assessment of these risk factors, an estimate of the probability of success can be determined for each stage of development. Most risk assessment processes produce numerical estimates of the probability of success at each phase. However, some approaches provide a qualitative estimate for the probability of success, such as categorizing drugs as having a high, medium, or low level of development risk.

The process of determining the probability of development success often starts with the identification of the key risk factors and then mapping the resolution of those risk factors to the different stages of a drug's development (Figure 8.6). The same categories of risk factors usually need to be considered for each drug and generally include efficacy, safety, pharmacodynamics and pharmacokinetics, formulation, and manufacturing, among many others. Within each category, risk factors may be broken down further, such as genotoxicity, carcinogenicity, dose range toxicity, and enzyme induction for safety. Individual risk factors can then be mapped against the phases of development when they are expected to be partially or completely resolved.

Having successfully mapped risk factors to different phases of development, the next stage is to analyze the potential outcomes and their likelihood for each risk factor at each stage of development. Based on the potential outcomes for different risk factors during a particular development phase, a probability of success is estimated. This may be based on the overall perception of risk, determined as the product of the outcome probabilities for individual risk factors, or on a risk benchmark adjusted for the perceived increase or decrease in risk. The output of this process is usually a numerical probability of success for each major development phase. This determines the probability of successfully reaching the market and also for failing at different phases of development.

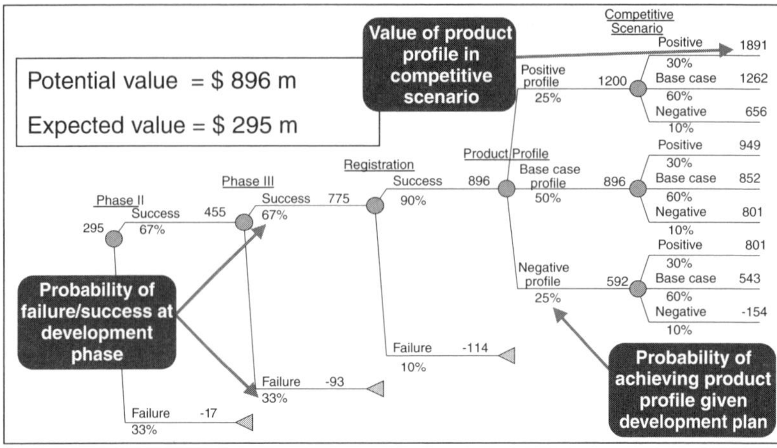

Figure 8.7 Value tree for a development investment.

Having determined the drug profiles that could be achieved and their commercial potential in different competitive environments, together with the cost and probability of successfully completing development, a valuation for a development project can be produced. The value of a development project is calculated as the cumulative sum of the probability-weighted NPV of the drug in each of the different potential investment outcomes. This can be visualized in the value tree for the investment (Figure 8.7).

The calculated eNPV for the drug represents the current average value of the drug across all the different potential outcomes. The range of value for the drug in different outcomes can be visualized using a risk profile diagram (Figure 8.8).

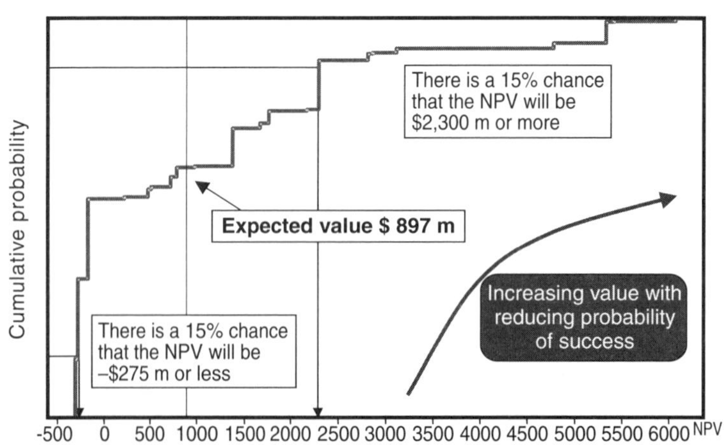

Figure 8.8 Risk profile diagram for the development investment.

The valuation of drug development investments is on the whole performed by valuation experts using inputs from functional specialists from marketing, regulatory, clinical, and manufacturing. Frequently, the valuation process centers on valuation workshops that explore and elicit the key valuation issues around the product profiles, commercial scenarios, development plans, and project risks. Outside of these workshops, individual experts will formulate the inputs to the valuation, such as sales forecasts, cost estimates, and probabilities of success. The inputs from specialists are collated and combined by valuation specialists. Challenging the opinions and expectations for an investment is a key component of the valuation process, improving the robustness and reliability of the valuation inputs.

A number of key learning points emerge from the approach taken for valuing development investment when attempting to value research investments. Most important is the structured process for valuing investments where the valuation is broken down into its constituent parts. Risk assessments, development plans, and commercial outcomes are segmented into the underlying issues and factors that can be more easily evaluated. Representative scenarios are used to capture the wide range of potential outcomes on different factors. Comparisons against existing therapies and needs of the marketplace provide benchmarks for assessing performance. Breaking the valuation down into smaller, more manageable parts makes it easier to produce answers to difficult valuation questions. A similar approach can be adopted for the valuation of research investments.

8.3.2 *Valuing Drug Discovery*

While most pharmaceutical companies perform valuation analysis on their development investments, relatively few evaluate research investments in the same way. This is largely driven by a belief that valuing research investments is too difficult and that the valuations produced are neither reliable nor robust.

However, investment decisions within research still need to be made and supported. The basis for making these decisions is more likely to be based on a general review of the strengths and weaknesses of an investment than on a financially oriented analysis of the potential return that may be achieved. In addition, obtaining a measure of the financial prospects for an investment is essential in order to justify an investment decision. If a company does not believe that a research investment will produce a financial benefit for the firm, why pursue the investment in the first place? The challenge is to elicit that reason. It may not always be possible to produce a precise monetary valuation for a research investment, but it is vital to acquire some indication as to how valuable an investment may be.

The quantitative approach for valuing development investments may not always be applicable to the evaluation of research investments. For instance, it might not be possible to determine a credible quantitative estimate of the probability of success or produce robust sales forecasts. If robust valuations can be produced using a quantitative approach, then this is the preferred approach, since it produces a simple measure that is easier to understand and compare to valuations for other investments. But it may be more realistic to use a qualitative approach that captures the general potential of an investment. However, the basic principles of measuring the potential reward of an investment and the likelihood of achieving that reward remain the same.

8.3.2.1 Understanding the Target Application

Most drug discovery projects focus on discovering drug candidates for specific disease targets using specific therapeutic approaches. This provides a defined-end application for the purpose of the research, giving a context for performing a valuation analysis. Understanding the possible end applications of the output from a research investment is the first step in the valuation process.

It is not sufficient simply to understand whether a disease or therapeutic area has a significant market value and is therefore worth researching. A careful assessment of the sources of met and unmet needs and the value patients, physicians, and payers place upon meeting them should be performed to understand what the key value drivers are for treatments within the disease area. This analysis provides the basis for evaluating the potential value a particular research project may attain by meeting different therapeutic needs. Producing such an analysis is critical not only for evaluating research investments, but also for devising a research strategy, assessing development investments, making technology investments, and selecting the best marketing strategy.

Analyzing the met and unmet needs in a disease area first requires the segmentation of the disease population into groups with common therapeutic needs. Though there will be many common unmet needs across a disease area, it is likely that there will be key differences in some therapeutic needs between different segments of the disease population. This creates different therapeutic preferences, and therefore value preferences in treating patients within each segment. Segmentation of a disease area should be based on key factors that produce clearly differentiated patient populations. For instance, segmentation may be based on disease state or severity, together with responsiveness and tolerance to existing therapies. The objective in segmenting the disease populations is to identify groups of patients where there is likely to be a similar value response to different therapeutic offerings, by patients, physicians, and payers. This makes it easier to produce a more accurate value assessment for therapies targeting a disease by assessing the value it achieves in meeting the different needs within each disease segment.

Having defined the disease segments for evaluation, the met and unmet needs across and within each disease segment need to be identified and their importance in treating the disease evaluated. This enables an assessment of the value placed on meeting those identified therapeutic needs. The identified needs will normally be centered on the efficacy, safety, convenience, and cost of treating the disease. These represent the key performance metrics for evaluating alternative therapies. Efficacy considerations may be on not only the required level of efficacy, but also the mechanism of action, pharmacodynamics and pharmacokinetics, and mortality and morbidity. There may be a critical efficacy need for a disease-modifying therapy for a disease where current therapies only alleviate or suppress the disease. Safety needs may be centered on toxicity, side effects, or adverse reactions. Convenience may be a critical issue within a disease area or segment. There could be a need for oral therapies where only IV therapies are currently available. Alternatively, there could be a need for slow-release therapies to improve therapy regimen and increase patient compliance. Therapeutic needs should be evaluated at a level specific enough to be clearly defined, but not so detailed that the analysis of therapeutic needs produces an unmanageable diversity of needs.

Having identified the therapeutic needs for each disease segment, it is necessary to understand the potential changes in treatment performance on those needs that create a step change in a therapy's value. A simple example would be the need for an oral therapy for a disease area where now only IV therapies are available. Significant value is likely to be placed on providing an oral therapy in this case. In contrast, developing a new oral therapy may not create any additional value in a disease area where that method of delivery is standard. However, value may be destroyed if a new therapy is not oral, since it would be less convenient than existing therapies.

Clearly, defining step changes in performance on different therapeutic needs provides not only the basis for establishing the value that can be achieved in a disease area, but also objectives and guidance in the research and development of new therapies. Considerations such as the convenience of a therapy and its cost of manufacture may not be at the forefront when evaluating a research investment. But they may actually play a far greater role in the value of that research than achieving a small improvement in efficacy.

The objective in evaluating the therapeutic needs for a disease is to provide a foundation for evaluating the commercial potential of different therapeutic approaches toward treating a disease, based on the perceived benefits and disadvantages of those approaches. Such an assessment means establishing an estimate of the value placed on alternative levels of performance on different therapeutic needs.

Alternative approaches exist for measuring the potential value associated with different levels of performance. At first glance, determining a monetary value for providing each level of performance on each therapeutic need may seem ideal. The value of a therapy could then be determined as a function of its expected performance on the different therapeutic needs. However, in reality, the sales and therefore value of a therapy are driven by a combination of its price and sales volume. Both price and volume are largely driven by the perceived overall performance of the therapy in comparison to alternative therapies. A small overall benefit for a therapy over its competitors may allow a therapy to capture a significant market share and therefore realize considerable value. Hence, it may not be feasible or practical to determine the monetary value of individual performance levels on different therapeutic needs, as the market does not view therapies in this way.

A better approach may be to obtain a judgment on the value importance of different performance levels in determining the overall performance for a therapy. To achieve this, a comparable measure needs to be used to assess performance levels across all needs so that the relative importance of different performances on different needs can be determined. A quantitative approach such as a scoring method or a qualitative approach such as assigning value levels (e.g., high, medium, or low) can be used. The challenge with both approaches is to obtain a robust judgment of the relative value placed upon different therapeutic needs and levels of performance on those therapeutic needs.

Formulating a robust judgment on therapeutic needs and performances requires the acquisition and careful consideration of any available market intelligence. This includes the opinions of thought leaders, surveys of physicians and patients, market studies, and trial results, together with the opinions of internal experts across all functions embracing research, development, regulatory, manufacturing, and marketing. Understanding the performance characteristic and perceived value of therapies available on the market, especially the current gold standard, is a key element in understanding the therapeutic needs. The expectations for therapies coming through the industries' R&D pipelines may also provide guidance.

Using a quantitative approach enables the overall value of a therapy to be calculated as a function of its performance on individual therapeutic needs, making the determination and comparison of the overall performance of different therapies easier. In contrast, a qualitative approach would require a judgment on the overall performance of a therapy based on a qualitative assessment of its performance on individual factors. This is likely to be difficult. People often find it challenging to make judgments based on a wide range of different factors that have fuzzy definitions.

The output from evaluating the target applications for a research investment is not only an estimate of the potential market size and value of the application, but also an assessment of the key value drivers for therapies in that market. This provides the foundation for assessing the potential benefits that a research investment may bring and the value it may attain.

8.3.2.2 Assessing the Commercial Potential

There may be many different therapeutic techniques treating a disease with many different disease pathways and targets. The challenge for researchers is to invest in the best therapeutic approaches for the right disease targets that will produce the best therapeutic results. To understand which approaches are likely to produce the best therapies, and therefore greatest value, requires an evaluation and comparison of competing therapeutic approaches.

In evaluating the commercial potential for a research investment, the expected therapeutic benefits and weaknesses need to be assessed and compared against the competition. A comparison needs to be performed against both currently marketed therapies and those that are still in research and development. The expected performance of a therapeutic approach should be assessed against the background of the identified therapeutic needs and desired performance levels within each disease segment. This can be achieved by mapping the performance expectations for a therapeutic approach against the previously determined therapeutic performance needs for a disease.

In assessing the expected performance of a therapeutic approach, emphasis needs to be placed on how the performance will be achieved. What benefits does the therapeutic approach have? The value placed on those performance needs by the market can then be used to determine the overall potential of a therapeutic approach. The same methodology can be repeated for alternative therapeutic approaches, enabling a comparison between competing therapeutic approaches to be more easily performed.

Predicting what the expected therapeutic performance of a polyclonal antibody approach is in comparison to a gene therapy approach for the treatment of lung cancer could be very challenging. This is particularly so when there is little experience or knowledge about either approach within a company. It may be impossible to predict or estimate the potential performance of a therapeutic approach on some needs. In circumstances where a prediction of the expected performance on a specific need or needs cannot be obtained, it should be highlighted as a risk that must be taken into consideration when assessing the probability of success.

The objective in determining the relative performance of competing therapies is to enable a prediction on the market success that a therapy may achieve within different disease segments. Based on the estimated value of a disease across individual disease segments, together with a prediction of the market success of a therapeutic approach given its expected performance, an estimate of the potential value of a therapeutic approach can be produced. Ideally, the estimated value will be specified in monetary terms, most likely as an estimate of the peak sales that could be achieved at today's prices. Alternatively, an estimate of the expected peak market share could be determined, which could then be used to determine a peak sales estimate. The established peak sales estimate can be combined with expectations around patent exclusivity and sales uptake to establish a potential value for the therapy. This produces a comparable measure of potential value for a research investment.

8.3.2.3 Evaluating Investment Risk

Establishing whether a research investment is likely to be successful is probably the single most critical element in the valuation of a research investment. Though a research investment may potentially lead to a new blockbuster therapy, the probability may be so low that the investment is not worthwhile. In contrast, a research investment targeting a low-value disease with a therapeutic approach that has a relatively high chance of success may have a greater expected value. Deciding which investment to pursue is then dependent upon the risk tolerance and value preference of the company.

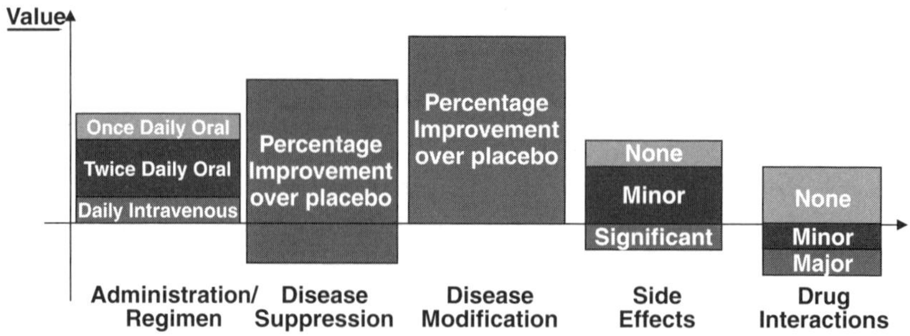

Figure 8.9 Relative value of therapeutic needs.

Determining the likelihood of success for a research investment requires an understanding of the potential risks for a therapeutic approach and the implications of alternative outcomes on those risks for its value. The first step is to identify the risk factors associated with a therapeutic approach. Potential outcomes for those risk factors can then be determined together with the likelihood of achieving different outcomes. Subsequently, the implications of expected risk outcomes for the overall success of the approach can be assessed. Finally, a judgment on the overall likelihood of success can be established based on the expectations for successful outcomes across all the identified risk factors.

Efficacy, safety, pharmacodynamics, pharmacokinetics, formulation, delivery, and cost represent the key categories of risk that are likely to influence the probability of success for a therapeutic approach. These largely mirror the factors considered when assessing the potential performance of a therapeutic approach. Though it is unlikely that one can foresee every possible risk for an approach, an attempt should be made to identify all the specific individual risk factors. Some, such as carcinogenicity, may lead to the eventual termination of a research program. Other risk factors, such as efficacy, may affect the performance or product profile of the therapy produced.

Where possible, potential outcomes for a therapeutic approach should be established on the identified risk factors. This may be limited to a positive or negative outcome; for example, there may or may not be carcinogenicity, or there may be a continuous range of potential outcomes such as different levels of efficacy. Where risk factors are associated with meeting identified therapeutic needs for a target disease, outcomes should relate to the desired performance levels for those therapeutic needs. For example, Figure 8.9 shows the estimated value of the therapeutic needs and of the possible outcomes of a research investment along these needs. Once the value of possible outcomes is assessed along key therapeutic needs associated with a research, a judgment can be obtained on the likelihood of achieving the desired level of therapeutic performance.

The implications of different individual risk outcomes on the overall success of a therapeutic approach need to be assessed. Risk outcomes primarily affect the success of a therapeutic approach by either causing the approach to fail completely or producing a relatively poor-quality therapy. Risks should be evaluated for both the likelihood of causing the approach to fail and also their expected impact upon the performance of the therapy, assuming it is successful. Uncertainty about achieving different levels of performance on identified therapeutic needs has a direct impact on the overall performance of a therapy, and therefore its potential value. A possible approach for handling this uncertainty is to develop representative product profiles for the potential performance of a therapy, assuming it is successful. This approach closely mirrors the approach taken when evaluating development projects. The product profiles defined should be differentiated by a limited number of key performance needs that drive the value for a therapy and around which there is significant uncertainty about the potential outcome for a therapeutic approach.

Risk assessments can be performed using either a quantitative, probability- based approach or qualitative, heuristic approach that defines risk in terms of perceived risk levels. Given the high level of uncertainty around research investments, it may be easier to apply a heuristic approach where the risk of different risk factors is defined using fuzzy risk levels such as high risk, medium risk, and low risk. Risk experts are likely to feel more comfortable providing a sense of direction for the risk on individual risk factors (e.g., high risk or low risk) than a specific number defining the probability of success. Subsequently, however, it may be preferable to translate the specified heuristic levels of risk into probability estimates using a standardized conversion guideline. Converting risk levels into probabilities makes it possible to calculate an overall probability of success for an approach across all risk factors considered. In contrast, attempting to convert a set of heuristic risk levels on different risk factors into an assessment of the overall risk associated with an investment is likely to prove substantially more difficult. Indeed, somebody's "high risk" may be somebody else's "medium risk." Obtaining a measure of the overall risk associated with a project in achieving its commercial potential is the key objective for the risk assessment process.

Establishing the level of risk on different risk factors for a therapeutic approach can be very challenging. There will often be "immeasurable" risk factors. In some circumstances it may not even be possible to identify them all. To overcome these difficulties, all available sources of information should be used. First, assessing the success of a therapeutic approach in other disease areas may provide some guidance as to the key risk factors and level of risk associated with a therapeutic approach. Similarly, understanding the risks commonly encountered with the target disease or therapy area may also provide some insight. It may be possible to develop risk benchmarks for specific types of therapeutic approaches or disease areas that can provide the foundation for risk assessment. If an estimate of risk cannot be established on some factors, then this should be indicated as a significant risk factor in itself, and the overall risk for the investment should be adjusted to reflect the high degree of uncertainty.

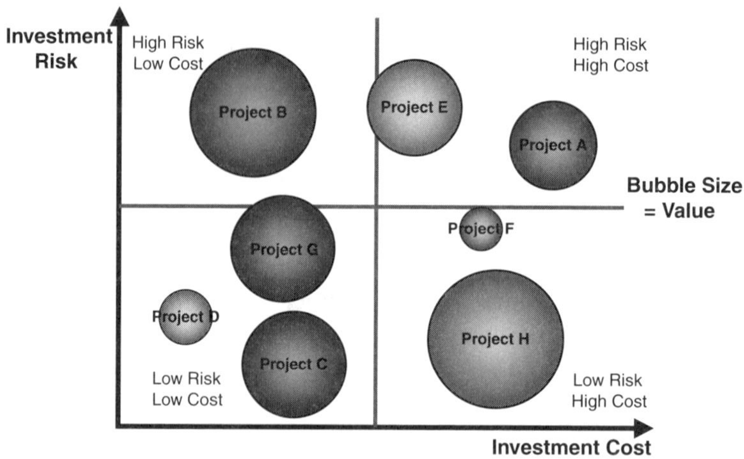

Figure 8.10 Visualizing research investment value, loss, and risk.

8.3.2.4 Determining the Expected Value

Having successfully estimated the potential value and expected risk associated with an investment, an overall risk-adjusted value can be determined. This represents the expected value from the investment and can be used in conjunction with the expected investment cost to assess whether the investment is worth pursuing on a financial basis.

Alternative research investment opportunities can be evaluated in a similar manner, enabling a comparison of their potential (Figure 8.10). This is essential to support the selection of research investments to be sponsored. In performing a valuation, the objective is not necessarily to obtain an accurate and precise value for the investment, but to generate a general sense of whether an investment is good or bad on its own, or by comparison to other investments in the portfolio. It is unrealistic to expect accurate valuations for early-stage research investments where there are so many uncertainties. However, without some formal processes for evaluating a research investment, there is a high probability of making false assumptions about its potential, and therefore of making a poor investment decision.

8.4 Making Your Valuation Process Work

The single greatest challenge to implementing a good valuation approach is obtaining buy-in from the organization, including both management and project teams. Without a strong buy-in from the organization, research valuations will prove too difficult to implement. To overcome these difficulties, the valuation approach adopted by an organization must be simple, transparent, and inclusive, produce feedback, and challenge and be owned by the organization.

Having a clear and transparent valuation methodology and a process that is easily understood by all stakeholders is essential. Complex black-box solutions where numbers are entered at one end and results pop out at the other should be avoided. Both management and project teams must understand the process by which an evaluation has been reached for it to be accepted. Similarly, stakeholders must understand how investment decisions are reached and the criteria upon which they are based.

Project teams hate providing information for investment evaluations without receiving any feedback as to the result of the evaluation, until their project is chopped. The evaluation process should help project teams in developing their investment plans to mitigate risks and increase value. Feedback must be an integral part of the investment evaluation process.

It is essential that project teams take ownership of project evaluations and drive the investment evaluation process. Ensuring project team ownership of investment evaluations increases the active and willing participation of project teams. This leads to more reliable and credible investment evaluations that are more likely to be accepted by the organization. At the same time, project teams generally prefer to be doing their primary job (i.e., researching new therapies), rather than spending time producing forecasts and risk assessments. Ensuring that the evaluation process is light and simple while producing pragmatic valuations is critical to achieving the willing and active participation of project teams. There is no real benefit in spending huge amounts of time and effort producing evaluations for projects that have only a relatively small value.

Investment evaluations should not represent the opinion of an individual person or function. The evaluation process should provide an opportunity for all stakeholders to voice their opinions. Bringing cross-functional groups together should help the identification of key risks and potential opportunities for an investment, not only helping to produce more robust evaluations, but also possibly increasing the value obtained. Similarly, investment evaluations should not be taken at face value. Management and project teams need to have confidence in the quality of investment evaluations for them to be relied upon. Ensuring that evaluations are challenged through peer review and compared across investments increases their robustness, credibility, and reliability.

8.5 Conclusion

Faced with a wide range of alternative research investments and a limited budget, life science companies have to choose the research investments they will pursue and those that they will drop. Valuations provide a meaningful and comparable measure of the potential financial reward that can be achieved through an investment, enabling a rational choice between alternative investments to be made based on the expected financial benefit. This provides life science companies with the basis for effectively allocating scarce research resources across their portfolio of investment opportunities to achieve their financial and research objectives.

Investment decisions are not limited to the sponsorship of drug discovery programs. Significant investments must also be made in discovery technologies and drug platforms and, increasingly, in external collaborations through licensing deals, mergers and acquisitions, and joint ventures— requiring a decision not only as to whether to make an investment, but also how much to pay for a company or what terms to offer in a licensing deal. Valuations help a company to understand the financial benefits and potential risks of an investment and the implications of alternative deal structures.

Traditionally, research valuations have been perceived as too difficult to perform, providing neither a reliable nor a meaningful measure of a research investment's potential. The key difficulty in performing research valuations is the uncertainty around the investment outcome, especially when focusing on discovery technology. However, without a robust measure of an investment's potential, little confidence can be placed in investment decisions that are made.

In developing an approach for valuing research investments, a great deal can be learned from the approach taken in valuing development investments. The approach we have taken for research investments is to focus on establishing a baseline valuation. To help us get there, we often break an evaluation down into the underlying factors that drive the value of an investment. In the case of research programs, emphasis is placed on understanding the unmet needs within a disease area and how a therapeutic approach might meet those needs. Combined with an assessment of the potential risk for a project, we can produce an estimate of the potential value, enabling a robust investment decision to be confidently taken.

chapter 9

Collaboration in a Virtual and Global Environment

John Barrett

Contents

9.1 What Do We Mean by Collaboration?

Collaboration seems to be a term that almost everyone in business is using but that currently has no consistent definition. For some, it means cooperation between departments where there was none or little in the past. For others, it implies sharing. Still others describe it as voluntarily working together. In a Gartner Group Commentary, collaboration is defined as "the process of working together toward a common purpose or goal in which the participants are committed and interdependent, with individual and collective accountability for the results of the collaboration, and each of the participants shares a common benefit" (1). While each of these may be accurate, the definition I choose to use in a business environment is people jointly working toward a common goal or objective. Of course "jointly working" deserves further clarification, so it should be understood to include freely sharing information and knowledge and clearly fulfilling responsibilities.

Today, the definition of collaboration also most certainly needs to include the tools that have been developed to support collaborative team activities. Collaboration work spaces are electronic tools that provide a configurable interface for consolidating team activities and information. Their features allow task or functional teams to easily share and manage content (documents and e-mails) and collaborate on other team activities using discussion threads, team calendaring, project scheduling, tasks, events, and change notifications. These work spaces can provide significant benefits to a team's productivity and effectiveness by:

- Reducing time spent looking for current documentation for information.
- Assisting team members to get up to speed more quickly.
- Capturing team history so it can be easily referenced in the future.
- Increasing the speed of communications and decision-making.
- Enhancing nonface-to-face collaboration.
- Simplifying working remotely or disconnectedly.

Finally, before moving deeper into our discussion of collaboration, I think it is important to position it relative to knowledge management (KM) because sometimes these terms are used as synonyms. If we think of KM broadly as the creation, capture, and leveraging of an organization's knowledge, then collaboration can be considered one of the ways that knowledge is created and leveraged. That is, by working together individuals are positioned to create knowledge, and if knowledge is shared while working together, then knowledge is being leveraged. So under the rubric of KM, collaborating can be thought of as a fundamental activity.

9.1.1 *Why Is Collaboration Important?*

In looking at my definition it would appear that some level of collaboration has been existent in business for a number of years. In fact, it could be argued that one reason firms are established is to facilitate collaboration among functions and activities that are required to produce a product or service. Of course, one could just as easily argue that the hierarchical organizational structure found in most businesses frequently hampers collaborative efforts.

So why today is collaboration one of the buzzwords in the business lexicon? First, quite simply, whereas in the past collaboration may have been a luxury or choice in the way an entity approached achieving its goals or objectives, today it is a necessity. As Ray Lane, former COO of Oracle and now a board member for MetaMatrix, said, "Collaboration is the most important word in business today." Gone are the times when businesses were so vertically integrated that collaboration across their supply chain was unnecessary, and likewise, the criticality of speed to market requires jointly working across functions to reduce cycle times.

Second, knowledge in many areas has become very specialized and deep. Often organizations are finding that their core capabilities, where they have special knowledge, need to be combined with capabilities provided by other organizations to be able to develop and bring innovative products or services to market. This combining of knowledge almost always requires collaboration. As Ross Dawson states in his book *Living Networks*, "You must bring together more and better resources than you can hope to have inside a single organization. This means that distributed innovation models must address how you attract the best people to collaborate with you in your projects" (2).

Third, the potential to easily find and connect or network with expertise in any location has become pervasive. No longer can geography be considered an intellectual competitive advantage or a detriment. So there are many more opportunities to work virtually collaboratively than ever before. As an example, a struggling Canadian gold mine posted its geographic data online "so they could attract the attention of [global] world-class talent to the problem of finding more gold." The winner was a firm from Australia who never visited the actual site (3). Virtual collaboration at its best!

9.1.2 *Most Teams Have Some Virtual Aspect*

Think back 10 years and most teams or projects had little if any virtual aspect. All or almost all members were likely to be located in close proximity of each other. For those not, then the only choice for working together was teleconference or travel.

It has all changed now. Many companies have become global or are partnering with others across the world. Even if all members are in the same geographic region, they are likely to be spread across different campuses, with the commute precluding frequent face-to-face meetings.

New technologies have provided the opportunity for separated members to participate virtually in almost all activities. However, we are still early in the development of collaboration techniques that promote being highly effective in a virtual environment.

9.1.3 Are Internal Collaborations Different from External Collaborations?

In many respects, whether all team members are from the same organization or not will make little difference. The "virtual" issues that will need to be faced are in most cases people related, not organization related.

However, there are two components where external collaborations present added complexity. The first, and maybe most obvious, is information security. The exchange of information outside a corporation's firewall requires appropriate due diligence, planning, rules, and technologies. In some cases, a collaboration tool used internally cannot be used by teams with external members because it does not meet an organization's security requirements.

Second, and often more problematic, is the fact that every organization has its own norms for team behavior and individual interaction. In most cases, these are so well established and known within an organization that they rarely need to be discussed. For example, if internally it is the custom to respond to a team member's request within 8 hours, then working with an external organization where the norm is 48 hours will certainly create problems. Indeed, different organizational norms can make collaboration with external members more difficult.

9.1.4 Everyone Has Been There

I would suspect that we have all participated on teams or projects where collaboration among the members should have taken place, but was lacking. I have witnessed any number of reasons, including lack of high-level sponsorship, functional objectives that were at cross-purposes with the team or project, and lack of expectations being established as to how collaboration should proceed. I can recall one situation where I never learned about information important to my tasks until the biweekly team meeting. Often many of these reasons exist concurrently, making successful collaboration a long shot indeed.

So what happens when there is a collaboration failure? At an individual level there is usually frustration that other members are not "doing their part" and that they are putting something else above the success of the team or project. At a project level, deliverables or activities are delayed or have to be reworked, resulting in timing or budget impacts. In addition, opportunities for identifying novel and innovate ideas will likely be missed. In the most severe cases, errors or omissions will occur that can have significant business impact.

9.2 Barriers and Obstacles to (Virtual) Collaboration

If the need to collaborate seems to be so important in today's business environment, why does it not happen more easily? Some of the obstacles relate to the teams themselves, where others are more rooted in what I call the organization's ecosystem. What is almost certain is that virtual teams will encounter team development issues that are not initially obvious.

9.2.1 Team Composition

Where as a few short years ago most of those working together toward a common goal were geographically located together, today that is certainly not the case. Being colocated does make working together easier and collaboration seem more natural. With today's dispersed (globally, telecommuting, multiple campuses) workforce, collaboration needs to be planned for and supported to succeed.

First, there is the basic issue of time zones. Synchronous collaboration is just plain difficult when team members are located across the world. While much team work is likely able to be accomplished in an asynchronous mode, certain types of issues or discussion content dictate everyone being together at the same time.

Different country cultures and languages can create issues. Even though English is the business language that is likely to be used, for those individuals for which English is not their native language, subtle nuances in a discussion can be difficult to pick up or can lead to misinterpretation. From a cultural perspective, some team members may be more reserved or less forthright and can be overrun by members who are culturally used to being more assertive or outspoken.

Communicating virtually is thus not the same as face-to-face communication, and decision making in a virtual environment has some inherent potential risks. In a recent article in the MIT *Sloan Management Review,* the authors suggest that a "seeing first" approach is best when many elements must be combined into creative solutions; commitment to those solutions is key and communication across boundaries is essential—for example, in new-product development, because it can surface differences better than analysis and can force a genuine consensus. *Seeing first* means literally creating a picture with others in order to see everyone's concerns (4). Often this is difficult to do in all but the most elaborate of virtual team work spaces.

Finally, not all people are naturally well suited to be virtual team participants. In some cases, these individuals are not comfortable working alone or remotely. I recall a team member telling me that he would not use the asynchronous discussion tool to "talk" to another member because they had always met face-to-face before. In other situations they might not have the discipline to stay focused when removed from the main team.

So the bottom line is that virtual team members cannot be treated the same as those on teams that operate strictly face-to-face.

9.2.2 *Team Membership Volatility*

We have entered the time of free agency where moving from job to job or even career to career is an expected norm. It is estimated that individuals entering the workforce today will change jobs nine times by the age of 32, and for those already well into their careers, time will be spent with at least five different companies (5).

And to compound the impact of the above, as baby boomers start reaching retirement age, some occupations could see as high as a 50% loss. Additionally, businesses are doing their share to exacerbate the problem as the result of continual upheaval caused by accelerated business reorganizations, mergers, or downsizing.

The net result is that team membership is more volatile than ever, with current members leaving and replacements often coming at a steady pace. The impact is threefold. First, knowledge known only to the individual exits when he or she leaves the team. Second, time is wasted while the replacement is getting up to speed. Third, and possibly not as obvious but maybe most important, the trust among the team members, which is so crucial for collaboration, needs to be developed with the new member.

9.2.3 *Organization Ecosystem Balance*

Virtual teams and their members exist within a corporate ecosystem that includes culture, values, and organizational structure, management systems (e.g., performance evaluation, compensation), and technology. These teams generally are performing as part of a process and are charged with completing specific tasks. If all elements of the ecosystem are well aligned, the team has the support necessary to successfully execute its phase in the process (Figure 9.1).

However, should one element be overemphasized at the expense of others, or if one element is changed without considering the possible adjustments that should be made to others, the team process will be negatively affected. In essence, the ecosystem becomes unbalanced (Figure 9.2). This figure shows the most frequent case, where technology is deployed to improve a process

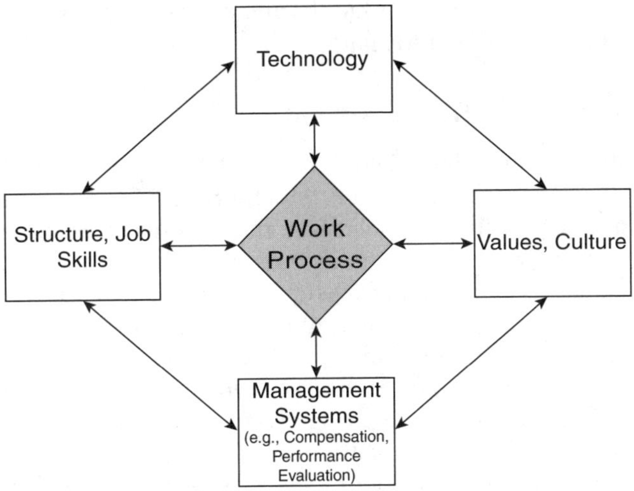

Figure 9.1 Balanced organizational ecosystem.

or to help the team without necessary consideration of the other elements. At best, the technology may have no impact on the team, and at worst, it negatively affects the team's ability to perform and collaborate.

9.2.4 Organizational Culture Support

One of the most significant obstacles that can be encountered is a culture that does not place high value on sharing of knowledge. This is often evidenced in the behavior of the leaders, lack of expectations being established

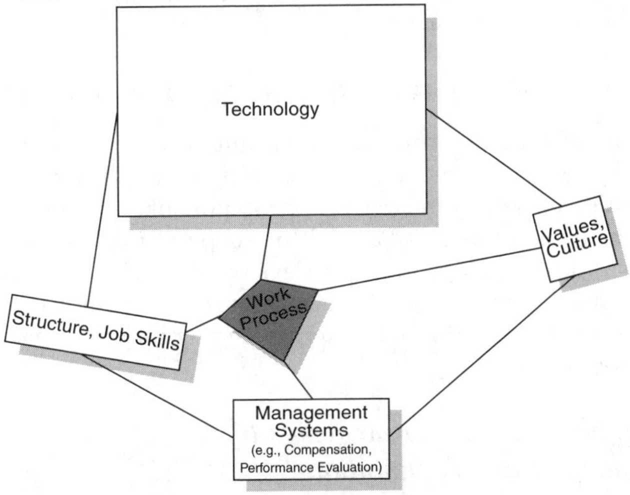

Figure 9.2 Unbalanced organizational ecosystem.

for collaboration, lack of knowledge sharing as a performance criterion, and lack of time being provided to share.

9.2.5 Not-Invented-Here Syndrome

Many scientists choose their field because of the opportunity to discover something new. A recent American Productivity and Quality Study on innovation found that "although scientists and engineers espouse values of knowledge sharing as central to innovation, there is a strong bias for invention and a reticence to reuse knowledge" (6). Failure to address this issue with specific approaches and interventions will result in severe limitations on the amount of collaboration that will take place.

There are a variety of approaches to overcome this strong cultural and professional bias, including facilitating diverse teams, making experts available to explain how an earlier invention could work in a new setting, rewarding for reuse, and storytelling about the successes from knowledge sharing, as well as a variety of methods to create relationships and trust across boundaries.

9.2.6 Research vs. Other Types of Teams

The question naturally arises as to whether collaboration and teaming are different in the discovery environment than, say, the clinical environment? It is often argued that discovery is inherently more entrepreneurial and creative than the process-intensive development phases. While this may be true at the core of discovery, with a researcher frequently working closely with a small team of assistants, today more than ever before there are many metaprocesses and project teams that surround the core research effort. Most all of these require the type of teaming and collaboration described in this chapter to be most effective.

9.3 Creating Value and Knowledge at a Distance

If we now understand that collaboration is important, but also that there are obstacles and barriers to success, how should an organization proceed? An implementation approach and necessary components are spelled out in later sections. However, while it is beyond the scope of this chapter to address how a collaboration strategy might be developed, it is important to list some of the principles, assumptions, and elements so the reader can see groundwork that must take place well before proceeding into the arena of collaborative environments.

9.3.1 Collaboration and Knowledge Management Principles and Assumptions

Since we are considering collaboration as a subset of knowledge management, many of the principles and assumptions that exist for KM apply

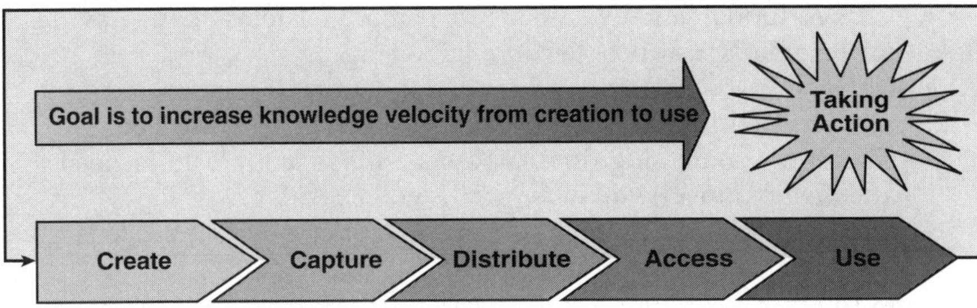

Figure 9.3 Speeding from knowledge creation to use.

equally as well to collaboration. As a first step, it is key to identify a set of principles that the strategy must satisfy and under which it will be developed and operate while helping achieve the business goals. While this may seem like a small undertaking, the results will have a profound impact on the strategy and its execution. The following are examples of principles for grounding a collaborative environment:

- It is a fundamental responsibility of all employees to actively share and seek knowledge.
- Knowledge creation, capture, sharing, and access will be built into performance and processes so that it is not an add-on, but part of the way we do business.
- Learning is the process or experience of gaining knowledge, and thus all learning will be integrated with the knowledge management environment.
- The speed from knowledge creation to use is paramount (Figure 9.3).

It is also critical to acknowledge certain assumptions and fundamentals that exist for all collaborative knowledge creating/sharing environments, including:

- Collaborative knowledge sharing (CKS) is rooted in behaviors: the key value necessary to support these behaviors is trust.
- The locus of all knowledge is the individual: individuals inherently want to know WIIFM (What's in it for me?) when asked to share their knowledge.
- CKS takes time and requires support: it will not occur if people are so stretched that they are already giving 110%.
- Knowledge can only be volunteered: we can always know more than we tell, and we will always tell more than we can write down (7).
- CKS requires demand in addition to supply; a build it (a repository), and they will come (approach) will not work. Sharing is demand driven, as we only know what we know when we need to know it (8).

- CKS is fundamentally person to person: systems may provide a starting point, but people share knowledge.
- CKS cannot be "bolted on": it will not happen if it is something that is positioned to occur after all the "real" work gets finished.
- CKS is not something done by others: those that create or need the knowledge must be involved in the collaborative sharing.
- CKS is formalized by process triggers and events: this embeds CKS into the fabric of the organization.
- Most knowledge-intensive business processes are complex: they cannot be broken down into a set of discrete elements, so opportunities for sharing and dialogue must be created.
- Knowledge is sticky: techniques are needed to lubricate the flow.

9.3.2 Collaboration and Knowledge Management Strategy Elements

A KM or collaboration strategy can be thought of as having three complementary elements: (1) architecture, the policies and processes that promote learning, seeking, collaboration, and knowledge exchange; (2) infrastructure, the tools that support effective access to and collaboration in creating knowledge; and (3) culture, the requirements and reinforcement that foster a shared sense of ownership and trust. All three elements—architecture, culture, and infrastructure—of the strategy must be complementary, which implies that their development is often very iterative. As one element moves forward, it generates the need to adjust the other two.

As mentioned in the opening section of this chapter, a collaboration environment may include multiple organizations. If this is the case, then discussions to develop a framework for evaluating requirements should occur soon after the strategy principles are initially identified, but before any substantive work on the strategy elements has commenced. This scenario would provide the basis for a more integrated set of actions than pursuing the elements independently.

After completing or having made good progress on the above steps, it becomes possible to start identifying the timing of tactics (projects and initiatives) to execute the strategy. This, of course, will then allow plans and budgets to start to be developed. A core set of resources will also need to be in place to lead and support the effort regardless of the specific tactics.

9.3.3 Critical Success Factors

Regardless of how good the principles are or how well designed the strategy is, there is a set of factors that will be critical to the success of implementing a collaborative environment. First and foremost, a high-level executive must express the commitment and desire to drive this. Second, and hand in hand with this, is the linking of a collaborative environment with a business objective, almost always in the high-level executive's area of responsibility. Third, is there an accurate assessment of where the behaviors currently are regarding sharing and collaborating? Fourth, is there a group or small area in the organization that is already collaborating and whose approach can be replicated? And finally, start small and test (pilot) before implementing across the entire organization.

9.3.4 Getting Started Fundamentals

If we assume that an organization has an adequate strategy now identified, how should it proceed with the implementation of a collaborative environment?

I suggest that the first step would be to conduct an assessment of where the organizations stands with regard to the culture, leadership, infrastructure, people, and process (CLIPP) characteristics necessary for supporting successful virtual team collaboration. Using some type of CLIPP survey (Table 9.1) tool will help identify areas for special focus and attention during implementation. Additionally, this should help to identify constraints and who/what else is or will be impacted.

The survey should also point out where education should be provided, which is normally the third step. Depending on the existing level of experience with collaboration in the organization, it may be necessary to provide education in the areas of team development, tool use, or even the basics as to why collaboration is necessary and beneficial.

The next step should be to select a team or two to work with in developing the details of the collaborative environment. These teams would not only have a role in providing input as to their needs and requirements, but would also be the ones to test the designs and solutions and then provide feedback. The key here is to select teams that are energized about contributing in addition to achieving their designated team objectives.

Once this design and testing step is completed, it is time to move into a pilot. A small but representative sample of the various types of teams that will be using the collaborative environment should be included. To provide a true indication of what can be expected during the enterprise rollout, every step of the expected implementation should be executed, all support elements should be tried, and all aspects of the solution should be used. It is during the pilot that marketing for the mobilization of the others who will eventually be affected should commence.

The final step would of course be the implementation for the entire selected population. Good project management techniques are essential for this to go smoothly, and plans should be such that early on those impacted will start to see tangible changes. In addition, there are some special considerations since the focus is collaboration. First, ensure that there is a means of sharing lessons and suggestions among the implementation team so that those that are implemented later can benefit from those who came on board earlier. Second, set up ways for teams to continue to share collaboration

Table 9.1 Virtual Teams: Assessing CLIPP Success Factors (9)

Culture	Firmly Disagree 1	Disagree 2	Neutral 3	Agree 4	Strongly Agree 5
1. Collaboration and teamwork are common.					
2. – 4.					
Leadership					
5. Leaders model collaboration behaviors.					
6. – 8.					
Infrastructure					
9. Adequate resources are provided to purchase and support collaboration technology.					
10. – 12.					
People					
13. Systems are in place to address career development for virtual team members.					
14. – 16.					
Processes					
17. Standard processes are used throughout the organization.					
18. – 20.					

Adapted from D Duarte, N Snyder, *Mastering Virtual Teams*, 2nd ed., John Wiley & Sons, New York, 2001.

ideas and suggestions after implementation of the environment has been completed.

9.3.5 *The Sweet Spot: People, Processes, and Tools*

While the potential benefits of collaboration are well known and have been noted earlier in this chapter, to increase the likelihood that, or the speed at which, the benefits occur, support in the form of guidelines, processes, practices, training, and reinforcement is almost always necessary. Thus, a successful collaborative environment must include components that address people, process, and tool aspects.

However, before dealing with the specifics in these areas, it is important to understand that there are aspects that transcend all three, which I consider meta-aspects. These include engaging executive leader sponsorship, promoting awareness of the need for and use of collaboration, showcasing recognition of the desired behaviors and actions, and collecting and documenting benefits or anecdotal success stories.

Maybe most important is ensuring that the individuals who need to collaborate have the capacity to do so. As Dr. Alex Broer, former vice chancellor of Cambridge University, was quoted as saying in *Working with Emotional Intelligence*, "You have to *talk* to everybody. So today you need more emotional intelligence than before to know how and from whom to get relevant ideas," let alone to form the coalitions and collaborations that will bring these ideas to fruition (10). Similarly, while noting in the book *In Good Company* "that collaboration—often involving numerous people in multiple locations—has become unquestionably necessary," the authors suggest that "the importance of NQ (network intelligence) or aptitude for connecting with others will be the glue that holds organizations together" (11). Thus, organizations need to recruit and hire employees and leaders with these capacities and provide coaching to develop them even further.

Indeed, even if all the items related to people, process, and technology are addressed, without the requisite attention paid to the proceeding meta-aspects, there will be little chance of successfully implementing a collaborative environment (Figure 9.4).

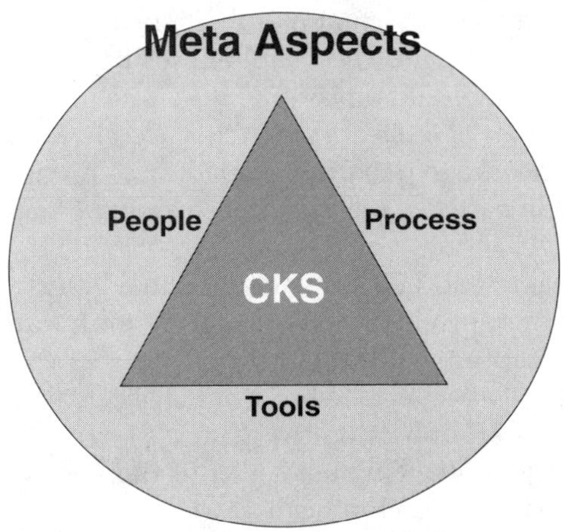

Figure 9.4 Collaborative knowledge sharing.

9.3.5.1 People

While collaboration by its very nature implies the involvement of people, it is amazing how often the implementation of collaborative environments is devoid of any focus on what is important in supporting those collaborating. Be assured that when this occurs, successful collaboration is left to chance.

First, and arguably most important, is defining and establishing the knowledge-seeking and -sharing behaviors that are crucial for team members to exhibit. This needs to include not only a list of the behaviors, but explicit activities that show evidence of the behaviors in action. The development of these must go hand in hand with working with the executive sponsors around defining their expectation for team collaboration and how they plan to reinforce it. The reinforcement element further expands the scope of this effort to include the human resources department, as it will be key to including these expectations and the recognition thereof into the organization's performance management systems. In the end, management leadership should decide and state how it plans to support and assess collaborative efforts.

Once the knowledge-seeking and -sharing behaviors have been defined, the second set of necessary people-focused activities is centered on the period of a team's start-up. This time is crucial to the speed with which the team becomes productive and to its long-term effectiveness. Team members must be clear on their respective roles and responsibilities, as well as the team's operating norms. Some of these can be captured in a team charter document (Table 9.2); others may best fit into a roles/responsibility matrix; a few such guidelines for communicating virtually will warrant their own documentation. However they are documented, they need to be created jointly by the team members, possibly as one activity during a team launch/building session.

Also during this start-up period, it will be useful to work with the team leader to help him or her understand how some facets of leadership are different on teams with virtual members. Leaders of teams with some virtual aspect need to expand their capabilities beyond those that make a traditional leader effective. For example, it is much more difficult to make everyone feel a full part of the team when they are always working remotely. How can leaders encourage rich multifaceted discussions when the technology often leads to linear exchanges or transactions? Other areas include:

- Understanding and being aware of virtual body language.
- Exhibiting high proficiency in virtual communications, sharing, and knowledge capture, and recognizing team members who also do so.
- Keeping members connected to what is happening outside the team, but may have an impact on the team.
- Sustaining and enhancing team energy and involving all team members in applauding each other's work.
- Identifying and refining ideas from team members.

Finally, soon after the team is up and running, another group of support activities needs to be considered. These include providing team tools for

Table 9.2 Team Charter

Name of team
Members
Meeting schedule
Purpose of team
Scope of project
Team shared values
Team norms
Roles and responsibilities
Team leader
Team members
Special roles
Team decision-making
Communications plans
Team internal communication
Team external communication
Grievances
Date/version/approval of charter
Charter expiration date/milestone

Note: The team charter contains the foundational information upon which the team is built. Agreement by the team on the information in the team charter is an important step in the development of the team.

areas such as decision making, conducting meetings, or reflection and critical thinking, and offering training on these and other team processes or tools. Ongoing informal assessment of the team's health should also be implemented and appropriate interventions at either a team or member level planned as necessary.

9.3.5.2 Processes

While teams that operate face-to-face can be somewhat effective without clearly defined team processes, this is less so the case with virtual teams. Virtual teams can get off track more quickly because the subtle recognition of and adjustments to performance deviations or behavioral anomalies that are almost intuitive when face-to-face do not occur.

9.3.5.2.1 Operational Processes. The first area of processes to be addressed deals with the everyday operation and activities of a team. These principally relate to how the team will use the work space to support its activities. Included should be recommendations and guidelines regarding virtual communications. What type of communications should be done virtually using work space tools, and when would virtual communications be potentially problematic, and thus should be replaced or augmented with more traditional synchronous modes? For example, virtual communications are not the best when delivering unpopular or bad news. On the other hand, capturing members' opinions about an issue or approach is generally well suited to polling or discussion tools.

A close corollary to guidelines about virtual communications is one that addresses e-mail. It should describe when a team might consider using e-mail for communications instead of the built-in work space tools. In general, it is usually beneficial to limit the use of e-mail so that all team conversations are captured in the work space and thus easily retrievable in the future. Normally, e-mail will be the best choice for communicating outside of the team. For those situations, guidelines for effective e-mail can be helpful and add a level of consistency for those who might receive e-mails from multiple team members. Additionally, these guidelines should indicate how this outside communication can be captured for later potential use and review by the team.

Other team activity processes where guidelines are necessary relate to the areas of content and document management. For content, the recommendations need to address what types of content should be stored in the work space and what should remain in other repositories and accessed by a link. For example, in most cases regulated content should not be placed in a work space but should be accessed via a link. While sometimes overlooked as a type of content, team-generated or received e-mails are often very important. The guidelines should address what types of e-mail should be posted to the work space and where.

Regarding document management, the guidelines should contain recommendations about the process for posting documents for comments, how drafts are to be handled, and when document versioning should be employed. For all types of content, team guidelines should be set for when it should be moved from the active areas of the work space to a historical archive structure.

9.3.5.2.2 Knowledge Management Processes. While less obvious, in many cases more important to a team's success are processes related to team knowledge capture. All too frequently teams are so task focused that they forget to reflect on what they have learned that could be of benefit to themselves or other teams in the future. This is a variation on the oft-quoted "Those who don't bother to learn from history are condemned to repeat it," by making the same mistakes or reinventing the same wheel.

Depending on the specific situation, there are a variety of knowledge-capture and -sharing processes that should be considered. First, if we think about the team composition volatility described earlier in this chapter, then processes for quickly "ramping up" new team members as they come on board, as well as those for capturing knowledge from exiting team members, are crucial. Both of these type processes must include techniques for dealing with tacit knowledge—knowledge that is not yet explicitly written down. Oft-quoted estimates suggest that more than half the knowledge that exists in a company is tacit knowledge. So good techniques such as storytelling and interviewing are necessary to codify and share the most important of it.

Another knowledge process that teams should have in their tool kits is one for capturing lessons and recommendations after they complete a specific phase or milestone of their business process or project. This is normally done through a facilitated lesson-capture session. A corollary process is one for capturing knowledge after events occur within a phase. This is commonly referred to as an after action review (AAR) and can be done expeditiously by those involved as soon as the event ends, using a template form. A third process that can be helpful is called the peer assist. With this the team requests input for addressing specific issues it is expecting to encounter from others who have already completed a similar project or process phase.

The three knowledge-related processes above can be combined into what is frequently called a lessons learned methodology (LLM). An LMM addresses how teams should learn before, during, and after the activities within a project or process phase. The most successful LLMs have project or process trigger points identified where the designated learning processes should be used. They also include prescribed means for other teams, communities of practice, or the organization to assess and take action as appropriate on the team's learnings and recommendations.

For both the operational and knowledge-sharing team processes, it is important to initially provide experienced assistance, as most teams (team members) are not naturally skilled or experienced in these areas. Additionally this outside team assistance can be invaluable in helping to define metrics for assessing the success or use of these processes.

9.3.5.3 Tools

Now that we have described many of the people- and process-related elements for successful team collaboration, we need to turn our attention to tools and technology. While not the most important element, these are enablers, or what can make the difference between collaboration and knowledge sharing being easy or difficult for team members. And of course, if it is difficult, then collaboration will occur less frequently and less effectively.

9.3.5.4 Electronic Collaboration Work Spaces

The most important tool is the electronic team work space. Collaboration work spaces such as Groove or eRoom are electronic tools that provide a configurable interface for consolidating team activities and information. Their features, although these do vary somewhat from product to product, generally allow task or functional teams to easily share and manage content (documents and e-mails) and collaborate on other team activities using discussion threads, team calendaring, project scheduling, tasks, events, and change notifications.

At a high level, the advantage of team work space tools is that they provide a focused location for all non-face-to-face team activities. In this regard, they are essentially a replacement for a physical team room or project office where most activity used to be concentrated when all team members where colocated. Thus, upon entering the work space, one is immersed in the team activities and sees or learns things that would not have been readily discernable if only, for example, e-mail was being used.

At a more granular level, the benefits of a work space tool can be mapped to a list of common team activities and tasks. This type of mapping should include how the team would or is performing the specified activities without the work space tool. Table 9.3 shows an example of a matrix mapping the features of eRoom and the task activities of drug project teams at a biopharmaceutical company.

If an organization has already selected a collaboration work space tool, then the first task is to learn about the team's activities before talking about the feature set of the work space. Using this approach allows benefits to be identified and keeps the focus on the team's objectives rather than the product.

However, if the tool has not yet been selected, how should one go about deciding what is the most appropriate work space tool? I suggest that the first two considerations are whether the team members need the ability to work disconnected from the network and if there will be members who are from organizations outside the IT security firewall. If the answers to both of these questions are yes, then the Groove platform should be at the top of the list of products to be considered. Don Courville, senior director of IT leadership at Pfizer, is especially appreciative of the mobility and cross-organizational aspects of Groove. He indicates that "they saved weeks during setup of spaces" with external team members because no inside firewall servers are required.

If most collaboration will be between team members within an organization and done while connected to the network, then a product like Documentum's eRoom may be the best choice, especially in the heavily regulated pharmaceutical industry. As Eric Miner, manager of knowledge management at Boehringer Ingelheim Pharmaceuticals, Inc., says:

> The manner in which it is implemented into an organization can be powerful. During template development, processes can be captured, gaps identified, and improvements made enabling an organization to coordinate, streamline, and innovate current processes and procedures. Decision making can be automated through the use of multiapproval databases and alerting functions. Folders and files can be incorporated such as to bring members of different locations and/or business units together into one shared working environment (i.e., medical and other departments can directly participate in discussions within R&D scientific eRooms). Thus, the eRoom solution can be utilized to break down both functional and geographic barriers as well as directly link business processes together with business objectives.

Table 9.3 Work Space Tool Support for Team Activities/Functions

Team Activity or Function	How Is It Done Today	Work Space Feature	How It Could Be Done	Potential Benefits
Member information	Each team member keeps or depends on the Project Manager (PM) to know	Database	Configure a team contact list database	No need for each member to make entries or be dependent on the PM With the new project–trial team linked structure there will be more associated team members Easier for new team members
Thought leader information				
Team meetings				
Events of interest to the team				
Storing team-created documents				
Document versioning and check-out/check-in				
Team e-mail archive				
Capturing process improvements and lessons learned ideas				
Real-time communication and presence indication				
Virtual voting				

(*continued*)

Table 9.3 Work Space Tool Support for Team Activities/Functions (*Continued*)

Team Activity or Function	How Is It Done Today	Work Space Feature	How It Could Be Done	Potential Benefits
Virtual discussions	E-mail forwarded	Discussion	Start a new discussion topic and send alert to members who should participate	Much easier to follow than forwarded e-mail Members can see most recent comments at any time and thus can jump back into the discussion at any time The thread is archived within the work space for historical reference and access by new members
Change in status notification				
Announcing new/updated information				
Task assignments				
Routing of documents				

Regardless of the work space tool selected, a number of items need to be addressed for it to be productively used by teams. In other words, despite what some vendors may say, teams are not ready to make effective use of them right "out of the box." First, you should work with individual teams to design and customize their space to make the best use of the tool features as related to their specific team activities and processes. These should be based on a set of design and configuration best practices. Second, guidelines need to be created for managing the work space. These should include getting started instructions and work space kickoff recommendations, as well as procedures for closing out a work space after is no longer needed. Finally, user support needs to be provided. In addition to standard help desk type support, this should include training, demos, and one-on-one special need assistance, as well as documented FAQs, tips, and instructions for more complex, less commonly used features. A listing of known issues is also usually helpful.

9.3.5.5 Other Tools

While the work space tool is the most important, some additional tools that assist in collaboration also deserve consideration. One of these is an expertise location management (ELM) system. Sometimes collaboration is limited because the others in the organization that could be involved or be contributors are unknown. This is particularly true in large globally dispersed organizations. ELM systems provide automated means for tracking and locating employees with expertise in various subjects.

A second helpful tool is a lessons learned or recommended practice repository. This can be invaluable in providing not just access to documented good ideas, but if properly designed, also the names of individuals who are knowledgeable about the ideas, thus enabling collaboration.

A third tool that may be considered is one that assists team members in global cultural orientation, familiarization, and education. If the team circumstances are such that the members come from different cultures and countries, and in particular this is their first exposure to the different cultures that will be represented on the team, then the opportunity for a cultural faux pas or lack of awareness that can slow team progress is great. A tool like Meridian Resources' "Global Teams Online" can then pay significant dividends in helping to avoid or resolve culture issues.

9.4 Case Study: Facilitating Collaboration on R&D Project Teams

Now that many of the details for creating a collaborative environment have been described, it is appropriate to describe how this plays out in a real situation. The following case study is based on an initiative at a midsize biopharmaceutical company to improve its compound development process by migrating from a functional orientation to one that is driven by matrixed cross-functional teams. It is called the high-performing team (HPT) initiative.

9.4.1 Background

The company, in existence for more than a decade, had recently been acquired by a much larger pharmaceutical firm. And as with most acquisitions, it was given aggressive growth targets. Although the company's serial, functionally oriented processes had served it well when it was smaller and when many of the employees had worked together for a considerable period, it was obvious it could not be scaled to meet the new targets. To compound the issue, the company was expanding from one campus to several with measurable commuting time between them.

Fortunately, although the new parent firm set aggressive growth targets, it also committed to support the resources and staffing necessary to meet the goals. The leader of the company's R&D group understood that to succeed in the HPT initiative would require formal processes and tools for the formation, development, operating support, and evaluation of teams and team members. Thus, he charted the R&D strategic team development group to support team development and team knowledge management and collaboration.

It should be noted that this was the company's first KM step, and it is not uncommon for companies to begin their foray into KM at a tactical level to address a specific issue or in support of another initiative like high-performance teams. While this tactical approach certainly provides benefits to the target initiative, it lacks the broad across-organization focus to have a major impact on business objectives if knowledge is truly a source of competitive advantage. To bring this about, a corporate KM strategy would need to be addressed.

In regard to this particular case study, it is interesting to note that no specific return on investment (ROI) was applied to the HPT initiative for spending resources to build a culture of HPTs. The leader of the R&D organization at the time seemed convinced that this was necessary to achieve the return he expected from his human assets.

9.4.2 Key Activities

The HPT initiative had two major components: team development and team knowledge management (Figure 9.5).

9.4.2.1 Team Development

The team development area focused on establishing behaviors and providing skills and techniques that would be required by team members to work effectively together. They began their work by surveying potential team members with regard to a number of teamwork-enabling factors. The results indicated, for example, that the team members understood the business reason for moving to the matrixed team environment, but were concerned that there were no tools for helping them collaborate, nor was there any means for capturing and sharing what they learned.

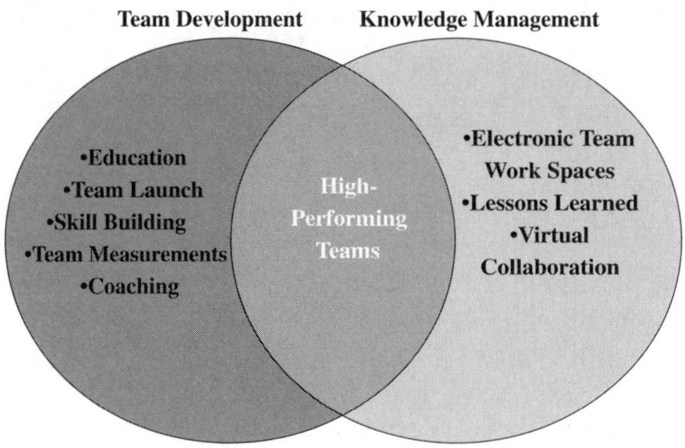

Figure 9.5 R&D strategic team development.

At this point a communications plan was put into place. Via posters, newsletters, and meetings with managers, the HPT initiative was described and implementation steps were discussed. The communications continued throughout the initiative and beyond, with team success stories eventually being highlighted.

One of the key steps was reaching agreement on behaviors, roles, and responsibilities for all team members and stakeholders. This listed how each level in the organization right up through the top-level governance board would interact with the teams and each other with regard to team-focused behaviors and responsibilities. As you may have suspected, knowledge-seeking and -sharing actions were a prominent part. Including the top level was significant and helped the rest of the organization to buy in, believing it was not a situation of "do as I say, not as I do."

To accelerate the move to effective teams, all potential team members were required to attend a workshop that would give them a common basis of team skills and behaviors. The workshop, titled Price of Admission (POA) (to participate on a matrixed team), included an overview of the new team model, assessment of personal preference and style indicators, description of roles, and discussion of teamwork competencies and behaviors.

Finally, as each team was formed, a team launch session was conducted. A common framework was created and then customized for each team as required. It included at a minimum creating a team charter, discussing team norms and expectations, and reinforcement of some of the content presented in the POA workshops. Often the teams also requested training or a demonstration of the collaboration work space tool, eRoom.

9.4.2.2 Team Work Space Implementation

The team knowledge management component focused on team collaboration and lessons learned. These were two of the items highlighted in the initial survey as needing additional attention.

eRoom had been made available in the organization more than 6 months prior to the HPT initiative being started. However, at that time there was no implementation planning done and it was offered with no business-related context, just a focus on the application. If not literally said, the position exposed was "Here's a tool with some neat features you can use, give it a try." As you might suspect, during that 6-month period, no more than a couple of teams tried it, and none were using it when the HPT initiative started.

After quick assessment of several teams to learn what benefits they required from a work space tool, a comparative analysis of eRoom and another tool that had some in-house sponsorship was conducted. eRoom was selected as the team work space tool because it met more of the required criteria and would be the least expensive to implement since it was already available on the corporate infrastructure.

It was decided to pilot eRoom with three different types of teams: a drug development project team, a clinical trial team, and a process improvement team. The specific teams for the pilot were selected based on their leader's willingness to participate and current phase of their project. The phase was important since we wanted the team to be relatively active during the 6- to 8-week pilot. As an example, one process improvement team was dropped from consideration when it learned of the need to compress its schedule and complete most of its work by sequestering themselves off-site for a period of more than a week.

As part of the introduction to each team, a matrix was prepared that listed many of the common team activities, such as maintaining team event calendars and managing document revisions. The matrix was then populated with information about how teams currently did that activity, how eRoom could assist with that activity, and finally the potential benefit to the teams. This was instrumental in creating a positive feeling about participating in the pilot and helping the teams to see how a work space tool could be used.

Prior to each team starting to use its work space, important discussions took place regarding team processes and practices so that the work space could be configured to best meet the teams' needs. As expected, the designs required some minor modifications as the pilot teams proceeded to use the work space, with the final configurations becoming the templates for constructing new work spaces for other teams of their type.

During the pilot period guidelines were developed for use by teams when the full implementation begun. These included work space design recommendations such as limiting the top level of the hierarchy to 15 objects or less. Recommendations regarding work space management included creating a table of contents when the number of documents in a folder exceeded 25. Guidance was also provided as to when a document should be stored in the work space and when it should be housed in some other repository and accessed via a link. General team practices were identified, such as responsibilities for updating team contact information or how to handle earlier drafts when a document is finalized.

Also during the pilot, a listing of tips and tricks was generated along with a small record of known technical issues. These were posted in a location that was easily accessible by all work space coordinators and were subsequently updated and added to as required.

The move from the three pilot teams to full implementation was seamless. As new high-performance teams were identified and staffed after the pilot, they were provided with their team work space. Initially, the plan had been to retrofit existing teams with work spaces at a steady measured pace. However, after individuals started working on a team that was using an eRoom, they quickly started clamoring for ones to be set up for the other teams they were on. The HPT group was only too glad to accommodate them.

9.4.3 Results

As previously indicated, there was no measurable ROI target specifically for the HPT initiative; it can be noted that the R&D organization did meet its objectives in 2002 where the teams had contributing responsibility. From the strategic team development group perspective, they have continued to see a strong level of requests for services and tools that can help teams perform even better, including decision making, facilitation, and meeting management skills. Looking at the work space implementation, within a year of the pilot there are more than 100 active eRooms supporting teams across the breadth of the R&D organization, including one used by the top-level management team to assist in its activities.

The following are quotes, obtained about 6 months after the pilot, from some team members about using the collaborative work space:

- "We are inundated with e-mail, so it's great to go to one place once to find information relating to the team."
- "With eRoom we never misplace any documents and are sure we all are working off the same version."
- "It is easy to find documents you need for your subteam or browse through other documents for learnings from other subteams."
- "We've used discussions to gain feedback on topics that need input prior to a meeting."
- "Document versioning was a big help when creating activity maps and modifying team charters."
- "Polls made it easy to find suitable dates for *ad hoc* meetings."
- "The alerting function is fantastic—you know when you can take action."

9.4.4 Lessons Learned

What are some of the lessons to be learned from this successful HPT initiative implementation? From an overall perspective, the involvement and sponsorship of executive management were crucial. Their willingness to exhibit the same knowledge-seeking and -sharing behaviors as expected of the teams, and being one of first to start using the work space tool, provided strong evidence of walking the talk.

With regard to the work space tool, describing its benefits in terms of team activities and providing easy access to expert user assistance helped teams overcome their reluctance to try it. Additionally, offering teams the choice of mode for instruction that best fit their needs—training classes, demos, or one-on-one sessions—allowed teams to more quickly acclimate to using the tool.

However, the key lesson is that facilitating collaboration requires a balanced focus on all organizational ecosystem elements (Figure 9.1). The proper combination of team development, knowledge management, and HR components prepared the way for success.

9.5 Summary

Collaboration, while important today, is rapidly becoming the way to business success. Organizations, teams, and the individuals who comprise them will need to excel at building and working in collaborative environments to be successful.

However, effective collaboration does not just happen, nor will it in most cases grow organically from the way that many organizations currently work. Processes and tools will be necessary, and significant attention must be devoted to human factors. The framework described above and the examples provided in the case study, if properly adapted for a particular situation, will go a long way toward enabling an organization to collaborate and create value and knowledge at a distance.

References

1. M Light, M Bell, M Halpern. What Is Collaboration? Gartner Research Note Commentary, COM-14-4302, December 21, 2001.
2. R Dawson. *Living Networks*. New York: Financial Times, Prentice Hall, 2002.
3. L Tischler. He struck gold on the net (really). *Fast Company Issue* 59:40–41, 2002.
4. H Mintzberg, F Westley. Decision making: it's not what you think. *Sloan Manage Rev* 42:89–93, 2001.
5. J Chatzky. Gen Xers Aren't Slackers After All. *Time*, Vol. 87, April 8, 2002.
6. P Leavitt. Using knowledge to drive innovation. American Productivity and Quality Center (APQC) 7, 2002.
7. D Snowden. Complex acts of knowing: paradox and descriptive self-awareness. *J Knowledge Manage* 6:103, 2002.
8. D Snowden. Complex acts of knowing: paradox and descriptive self-awareness. *J Knowledge Manage* 6:103, 2002.
9. D Duarte, N Snyder. *Mastering Virtual Teams*, 2nd ed. New York: John Wiley & Sons, 2001.
10. D Goleman. *Working with Emotional Intelligence*. New York: Bantam Books, 1998, p. 101.
11. L Prusak, D Cohen. *In Good Company: How Social Capital Makes Organizations Work*. Boston: Harvard Business School Press, 2001, p. 37.

chapter 10

From Genome to Drug: Ethical Issues

Patricia Deverka and David Magnus

Contents

10.1 Introduction

When posed with the question "Why should the biopharmaceutical industry concern itself with bioethics?" there are at least three compelling answers. First, these companies are investing heavily in genomics and proteomics for their potential to positively transform both drug discovery and drug development. Rightly or wrongly, DNA-based research is evaluated very differently from other types of traditional drug research. Advances in biotechnology are viewed as having the potential to shape the very nature of human life because they manipulate DNA, the building blocks of life. Descriptions by scientists of the importance of mapping the human genome have led the public to expect a transformation in biological research and medical practice (1,2). Characterization of the human genome sequence as the instruction book for human biology (3) only reinforces the notion that biotechnology overlaps with fundamental questions about what it means to be human.

Innovations in biotechnology have the potential to not only improve our health status, but also change how we live, what we value, and who we are—leading to debates about the moral dimensions of personhood (4). Therefore, it is not surprising that a wide range of stakeholder groups want a role in determining how biotechnology should be used, regulated, and financed, particularly as they have argued that the current oversight mechanisms are inadequate to meet the new challenges. Despite their specialized technical expertise, scientists and businesspersons are not viewed as having the requisite skills to address questions that are inevitably normative, value laden, and metaphysical in character. While many patients eagerly await the benefits of new and more effective treatments, other members of the public fear that we will simultaneously alter human nature in deleterious ways. While much of the current public debate about genetic advances has focused on stem cell research and reproductive cloning, in the near term a far greater number of people will be impacted by the new biotech drugs produced by scientists working in molecular biology and drug development (5). The role of bioethicists in this setting is to raise doubts and ask difficult questions about the morality of certain technological innovations, while simultaneously providing a justificatory framework for determining what actions are morally permissible and why. The objective is to help decision makers in the biopharmaceutical industry anticipate the potential outcomes of their R&D investments and develop ethically sound strategies for their research and marketing efforts.

The second reason for proactively addressing the ethical challenges of the *-omics* (genomics, proteomics, metabolomics, and so on) is that research in these subfields of molecular biology sits squarely at the intersection between science and technology (6). Molecular biology is an area of intense public and private research investment, which leads to a blurring of the historical separation of basic science and applied technology. Today, the biopharmaceutical industry is a growing source of funding for academic research in these areas and universities are encouraged to patent their discoveries. Both academic and industry scientists are working to develop the fundamental knowledge to improve human health and quality of life, as well as to translate that knowledge into new products and services. As in most situations where there is a rapid change in traditional roles and norms, the new rule set is still being worked out. While the goal is to develop policies that simultaneously preserve the benefits of academic biomedical research and proprietary product development, the overlap zone presents numerous ethical challenges because of conflicting values. An example is the ongoing intellectual property debate regarding the ethics of patenting DNA. Despite the established trend of granting DNA patents by the U.S. Patent and Trademark Office (USPTO) and its international counterparts, many members of the public and experts in medicine, law, and ethics continue to question the morality of these patents, particularly those covering research tools and diagnostic tests. The global biopharmaceutical industry must recognize both the opportunities and responsibilities that come with working in this overlap area. Again, bioethics has an important seat at the table where decisions are made about complex issues such as gene patents, research integrity, genetic privacy, and access to new drugs.

Finally, biotechnology represents an area of scientific inquiry where making a mistake is costly and can cause long-term negative repercussions for the field. The mistake in this case refers to not recognizing the broader ethical and societal implications of biotech products and failing to have a plan for introducing the innovative product that addresses these concerns. While not a new drug, the best example of how costly this oversight can be is the case of genetically modified (GM) foods, specifically the widespread negative reaction of consumers to the introduction of Bt corn, a type of corn that was transgenically altered by scientists working for Monsanto to make it resistant to pests such as the corn borer. Monsanto's mistake was that it thought of Bt corn simply as a new and improved agricultural product with the farmer as the target customer (7). When antibiotechnology consumer groups got wind of the new product, they protested loudly. Monsanto further compounded its error by extolling the virtues of GM food and was quickly characterized as arrogant and insensitive. Since that time, Monsanto has taken steps to change its image as the leader of the "frankenfoods" movement, but its reputation has been seriously damaged. More importantly for the field in general, the other companies working on GM foods were painted with a similar paintbrush. "Of the many lessons to be learned from Monsanto's approach, this is one that should not be forgotten: negative public perception can extend beyond a single company to any and all firms that develop the scrutinized technology" (8).

This case is an example of the tension that can develop when biotechnology companies and scientists believe that consumer objections to biotech products are rooted in the public's lack of understanding of the scientific facts. Therefore, they often advocate greater education about biotech products, both to dispel the public's (irrational) fears and to enable consumers to make informed choices. What the experts fail to appreciate is that disagreements about the appropriate applications of biotechnology are often not simply due to a lack of understanding of factual matters, but rather reflect a fundamental conflict in values. One important goal of bioethics is to bring the discussion of values to the forefront and provide a set of tools for weighing and balancing the competing interests in an objective and transparent manner. Thus, it follows that scientists and senior managers of biopharmaceutical companies must work together with bioethicists to anticipate the likely ethical issues and proactively develop strategies to avoid a public backlash based on misunderstandings and a conflict in values. This is one critical step that must occur to ensure that the benefits of genetically based new drugs reach the patients that need them, while directly addressing dissenting opinions in a manner that builds trust and support of the outcome.

10.2 Bioprospecting and Royalties

There are a series of important lessons that can be learned from the field of agricultural biotechnology, particularly as these issues play out on the international stage. For example, there are a number of concerns raised by the patenting of living organisms and germ plasm. These range from the metaphysical (what counts as part of nature vs. invention) to practical policy concerns. Among the primary considerations are worries over justice, particularly with respect to "bioprospecting" or "biopiracy"—the development and patenting of material derived from resources and knowledge in less developed nations for the benefit of corporations based in developed nations.

Until fairly recently, patenting practices varied widely from country to country, and practices that were common in the U.S. and other northern nations were often restricted or forbidden in the southern less developed countries (LDCs). The U.S. and European Union (EU) were often frustrated in their attempts to enforce intellectual property (IP) rights in LDCs. Many of these nations only allowed process patents, not product patents. Therefore, generic companies operating in these countries only had to produce the same pharmaceutical product in a different way to circumvent the patent a company had on a drug. Different countries also recognized patents for varying lengths of time. India, for example, only recognized process patents, and for only a period of 5 years. Many of the northern pharmaceutical companies found this a problematic situation for their research and development efforts. Pfizer found that prior to government approval of the antiarthritic drug Feldene, a generic competitor already existed in Argentina, and by the time Feldene went to market, Pfizer faced competition from six generic drug companies. Attempts to change the system to make it more uniform across national boundaries met with failure. The World Intellectual Property Organization (WIPO) was set up as a United Nations agency in 1967 to administer international agreements and treaties with respect to IP issues. The Paris Convention of 1883 required that each nation grant the same patent protection to people from other countries that it grants to its citizens. This requirement did nothing to stop countries from having IP systems that differed markedly from the U.S. system, as long as they were consistent. Therefore, Pfizer and other companies began to lobby the WIPO to change the Paris Convention (9). This effort met with failure. Later, the U.S., the EU, and Japan agreed to pursue an alternative. There began to be an increasing connection between IP and trade. This resulted in increasing pressure from the U.S. and other northern nations on the LDCs to comply with their patent protection systems or face trade sanctions. And through negotiations over the General Agreement on Tariffs and Trade (GATT) there emerged a "floor" governing IP systems in all GATT nations, through the Trade Related Aspects of Intellectual Property Rights (TRIPS).

10.2.1 Ethical Issues

The question that the current system raises is whether it is fundamentally unjust. LDCs are systematically disadvantaged relative to the interests of the U.S., the EU, and their multinational corporations. The traditional knowledge and the germ plasm of LDCs are mined for their value for industrial interests, often with little or no payback to the original developers of material. This is sometimes referred to as biopiracy. There are at least two different arguments. First, LDCs may be responsible for both the creation and preservation of valuable germ plasm. That these organisms have often resulted from years of agricultural practices (similar in many respects to scientific plant production (10)) and efforts by indigenous groups to preserve valuable and rare resources would seem to entitle the developers to some of the benefits that may accrue from the results of usage of the organisms or genes that they have helped to create and preserve.

Second, traditional knowledge also involves knowledge of how the raw materials can be harnessed for various purposes: medicinal, agricultural, and so forth. The argument that is made here is either that IP built on the basis of traditional knowledge should entitle the communities that created that knowledge with a share of the benefits that ensue or, more typically, that the traditional knowledge constitutes prior art and thus invalidates any IP claims. These two distinct aspects of biopiracy may be thought of as resource and knowledge biopiracy, respectively. It is important to recognize that these two arguments may conflict—one is aimed primarily at a share of the benefits of the products that eventually result, while the other attempts to invalidate the legitimacy of the patent claims, leaving development outside of the IP system. These arguments also lead to two very different legal strategies. One aims to invalidate a patent claim on the grounds of prior use, while the second acknowledges the legitimacy of the patent claim but makes the case that LDCs have a right to share in the downstream benefits.

There are several ethical problems with the current IP system as it stands. First, with respect to germ plasm from LDCs, critics have pointed out the disparity between the way genetic resources and other natural resources are treated. Petroleum or mineral resources are the property of the nation within which they reside. Genetic resources, in contrast, "have long been considered a common heritage available to other nations for free" (11). LDCs often act as stewards of these resources and even utilize scientific methods to develop them. But there is no systematic way of rewarding them for preserving and developing these resources. One solution to this problem has been to create the Convention on Biological Diversity (CBD). This requires that researchers from signatory nations seek permission prior to utilizing genetic resources from other countries. However, given the extent to which valuable genes and organisms have already flowed from LDCs to the north, and the additional problems of enforcement, it is not clear that the CBD is a panacea for these problems. Further, it is not clear how to address the conflict between CBD and TRIPS (12). From a moral point of view, the fact that local communities preserve and create genes and organisms would seem to require that ethically they are entitled to some form of benefit sharing in the fruits of future product developments based on their material. The CBD is one mechanism that can help promote benefit sharing, but more needs to be done.

With respect to traditional knowledge, the current regulatory system appears stacked against LDCs. For example, a U.S. patent can only be invalidated by claims of prior use if there is evidence of use in a form recognizable by the U.S. courts and PTO examiners. That is, accessible documentation or prior patents are needed to invalidate a claim. If indigenous, traditional knowledge is largely expressed in custom, habit, and oral traditions, it will not be recognized in the patent system. There may even be problems if the documentation is not in English or an easily accessible language. This results in systematically favoring nations with well-established IP systems similar to the one in the U.S., and it favors nations that have strong written, rather than oral, traditions.

In the area of biotech patents, particularly in the U.S., there seems to be a great deal of leniency in granting patents by examiners and reliance on the courts and other systems of appeal to overturn patents that are invalid. The problem with this is that it favors nations and corporations with the financial resources, who have an incentive to attempt to procure as many patents as possible, whether valid or not, and places the burden on poor LDCs or nongovernmental organizations (NGOs) to attempt to invalidate them. It is simply not possible for every one of the many alleged cases of biopiracy to be challenged by the LDCs or the NGO community. It is important to note that the previous argument (which applies to resource biopiracy) leads to a very different conclusion than this argument (which applies primarily to knowledge biopiracy). This argument challenges the validity of the patents at all, rather than making a case for benefit sharing.

A third argument against the current system is that the LDCs are arguably deprived of future potential benefits through the opportunity to develop products based on their indigenous genetic resources and traditional knowledge. This pits the interests of developed nations, which want products to be developed as quickly as possible, against the interests of the LDCs. Again, this system will favor those nations and institutions (corporations, universities) with the resources to develop products as quickly as possible, building on the genetic resources and traditional knowledge of LDCs (13). This is particularly problematic because in addition to causing the loss of future economic opportunities, it may undercut current local businesses as new products undercut existing ones.

Finally, we need to be concerned about the impact of bioprospecting and biopiracy on the lives of the people in LDCs, particularly when they result in profound changes in the local economy and traditional practices. Consideration of these impacts, for better and for worse, must be thought about as part of the consideration of the impact of bioprospecting. If the net harms in a particular case outweigh the benefits (for the LDC), there must be mechanisms that can protect the use of their germ plasm, their knowledge, so that it does not harm them.

Justice requires attention to more than the technical elements of current patent law. It is much more than a technical area of law, and the assessment of what institutions, laws, and practices we should adopt has to involve more than a scientific assessment of the risks and benefits of various IP regimes. Questions of who benefits, who has a claim on that benefit, who is exposed to risk, and who decides the allocation of risk and benefit are all moral questions that require much more serious attention. Since 1992, many LDCs began setting up systems for benefit sharing as a result of the Convention on Biological Diversity. For example, Novartis entered into a contract with Bioamazonia, the Brazilian government's institution that was created to manage the country's genetic resources. Under terms of the agreement, Brazil has received funding and training: the greater the number of marketable products, the greater the amount of funding. Novartis received exclusive access to the biological material of Bioamazonia.

10.2.2 Recommendations

These arguments underscore the importance of developing ways of sharing the benefits with the communities that materially contribute to the development of valuable IP, not just those eventually defined as the inventors (14). The ramping up of the IP systems in some of the LDCs might be helpful in trying to deal with some of these problems (15). Peter Drahos has suggested creating a global biocollecting society to better reward the contributions of indigenous groups for their knowledge (16).

Groups such as the Ethics Committee of the Human Genome Organization (HUGO) have published specific recommendations about how biopharmaceutical companies should operationalize the notion of benefit sharing (17). In recognition of the contribution of indigenous groups to genetic research, and based on the principle that the human genome is part of the common heritage of humanity, HUGO recommends that corporations donate 1 to 3% of their annual net profit to meeting the health care needs of these groups (18). Since there is the potential for a lot of money to be made from genetic research, the perception is that by doing nothing, biopharmaceutical companies are simply exploiting the people and countries that contribute DNA samples to genetic research. Benefit-sharing schemes are an essential strategy to avoid exacerbating the current economic and health disparities that exist between rich and poor countries, and reflect the special moral obligations that exist for companies working in human health (19). This notion that biopharmaceutical companies have specific ethical responsibilities to bring their resources to bear on alleviating human suffering, even when those efforts are not always profitable, is not new (20). What is new is the extrapolation of this obligation to genetic research, specifically in the form of benefit claiming by groups (such as a country or advocacy group) that negotiate for a piece of the economic pie by limiting access to their genetic material through the creation of a DNA registry or data bank (21).

Biopharmaceutical companies are not the only participants that have obligations in genetic research. Although bioethicists typically emphasize the importance of respecting the rights of the individuals participating in research, there are also obligations of the individual to share information (in this case, his DNA) that will benefit the group (e.g., a group with the same genetic disorder). This emphasis on the duty of individuals to participate in research for the common group is better captured in the ethical framework of solidarity (22), which is a very different approach to the notion of sharing the benefits of research from the one described above. A balance of both individual and corporate contributions will be required to ensure that considerations of justice and equity are not overlooked.

10.3 Gene Patents

Despite ongoing debates about the fairness of applying the current IP framework to genetic information, the practice of patenting DNA is well established, particularly in the U.S. (23). Since the landmark case *Diamond v. Chakrabarty*, where the U.S. Supreme Court ruled in 1980 that a biological organism (an oil-eating bacterium in this case) could be patented (24), there has been an explosion in the number of DNA-based patents. The Chakrabarty decision was viewed as a departure from the previously held view that it was not the intent of Congress to permit the patenting of life-forms other than new varieties of plants (covered by the 1930 Plant Protection Act for plants that reproduce asexually and the 1970 Plant Variety Protection Act for plants that reproduce sexually) (25). Following the Chakrabarty decision, patent office policy shifted to include "non-naturally occurring non-human multicellular living organisms, including animals" as patentable subject matter, leading to the first patent on a multicellular organism (the Harvard oncomouse) in 1988 (26).

John Moore was a patient who sued his physician and a biotechnology company after discovering that they had developed a cell line for commercial purposes based on his tissue biopsies without his consent. While recognizing that the defendants should have first obtained Moore's consent, the California Supreme Court ruled against Moore's claim to property rights over his tissues (27). The judges' rationale included the concern that recognition of an individual's financial interest in his own cells would impede medical progress by removing the economic incentives for companies to conduct important medical research (28). These are only a few examples of the types of DNA-based patents that are viewed as critical to launching the U.S. biotechnology industry (29).

10.3.1 Ethical Issues

From an ethical perspective, the most controversial DNA-based patents are those claiming human genetic material. The patented material can be a gene, DNA sequence, complementary DNA (cDNA), expressed sequence tag (EST), or single-nucleotide polymorphism (SNP) (30), referred to here loosely as gene patents for the purposes of discussion. The objections to human gene patenting fall generally into two broad categories: metaphysical concerns and concerns regarding justice and the potential for an undesirable imbalance between public and private interests.

For many individuals, patenting the DNA building blocks of human life appears wrong at face value. Unlike the genomes of other species, human DNA has a special status that merits differential treatment. This view is captured in the notion that the human genome is a shared resource that represents, in a symbolic sense, the common heritage of humankind and should not be used in its natural state for purposes of financial gain (31). Since we share 99.99% of our DNA in common with other human beings, no one group or company should be able to profit by monopolizing any component of our common heritage. Several oversight groups in Europe (32) have tried to codify their general unease with patenting human DNA, but the patent prohibitions typically only cover genes in their natural state. However, since patents are only granted for the isolated or purified forms of genetic material (not as they exist in the human body) (33), this position does not give real guidance for how to deal with the current state of gene patenting. Moreover, there is no legal recognition for the role of international stewardship of the human genome, leaving this "common heritage" objection irresolvable in a practical sense (34). Finally, the key justification for a patent system is to promote the general good, through encouragement of both investment (in research and development) and disclosure. Thus, even if the human genome is part of our common heritage, there is no reason why patent protection or other IP regimes could not be enacted as the best means of utilizing that heritage for the general good.

A related objection is that the practice of gene patenting will lead to a commodification of the human experience (35). Commodification describes the process of applying market-based values such as property rights to human bodily materials or even human beings, with the undesirable effect of reducing human life to an expression of monetary worth (36). Patenting of living things requires that we conceive of them in market terms. It implies both ownership and an instrumentalism that is incompatible with many views about the nature of life. This leads to a slippery slope that changes the nature of human interactions and threatens to undermine human flourishing (37). Proponents of this argument recommend severely limiting human gene patenting even in the absence of evidence of actual harms, because of the potential for an erosion of human dignity.

This objection has particularly been made from a theological perspective, as a gift from God (life) is transformed into an "invention" to be owned and used. It shows both a lack of respect and hubris with respect to the world and our relationship to it. This is often expressed in visceral terms as a concern over "playing God." Though this objection has some weight, it is counterbalanced by arguments about the benefits of utilizing the patent system as a way of encouraging the development of products and even organisms that will be helpful to humanity. Every religious tradition recognizes the importance of balancing the needs to "preserve the garden" and to "tend the garden." The key question is whether patenting of organisms, genes, cells, and so forth, pushes the appropriate balance to an excessively instrumentalist view of nature.

Finally, there has been an extensive debate over whether genetic sequences are really a product of nature, and therefore not eligible for patenting. Provocative questions such as "Who owns your genes?" fuel the controversy over whether genes meet the criteria for patenting, since in this view they already exist in nature and can therefore only be discovered, not invented by scientists (38). Practically speaking, this issue has been resolved in the courts, which in the U.S. have consistently ruled that genetic material meets the criteria for patenting as long as it is isolated from its natural environment and purified (39). These decisions rest on the distinction between owning the tangible property (the gene in your body, which cannot be commercialized) and owning the intellectual property (the cDNA, which can be commercialized). Currently, in both the U.S. and Europe, isolated DNA sequences are considered to be eligible for patenting, thereby neutralizing the product of nature argument from a legal perspective. Nevertheless, it remains a matter of ethical discourse as to whether the current patent system *unfairly* rewards genetic researchers because they have not done any real work beyond describing a natural occurrence (40).

The question of whether applying the standard criteria for patentability to genetic material is both fair and likely to produce outcomes that are just for society represents the cornerstone for the remaining series of concerns about gene patenting. The goal of the patent system is to benefit society through the disclosure of information about a new invention. In the case of gene patents, the public stands to benefit from this disclosure in terms of securing access to new DNA-based medicines, products, and services (41). To encourage disclosure by inventors, the government enforces a time-limited monopoly where only the inventor can realize financial rewards from the invention (42). In order to receive a patent from the government, the inventor must demonstrate that his or her invention is new, nonobvious, useful, and does not exist in nature in the same form (43).

Patents also allow investors to take risks more willingly and are essential to capital formation for unproven technologies. Corporations would not make investments in projects with high failure rates such as pharmaceuticals unless they were guaranteed the exclusive right to recoup their investment, albeit for a limited time (44). Therefore, although patents are not the only means of stimulating investment in research and development, they are viewed as mandatory by the biopharmaceutical industry (45). In return for patent protections, the pharmaceutical industry has delivered a broad range of new medications, many of which represent major treatment advances. For example, of the 192 new chemical entities approved by the U.S. Food and Drug Administration (FDA) between 1981 and 1990, 92% were developed by the pharmaceutical industry (46). Similar outputs in terms of new medicines for patients are beginning to be observed from biotechnology companies, with over 95 new biotech products on the market and over 370 in the pipeline (47). Patents on genetically engineered protein drugs, such as insulin and erythropoietin, have not been met with much controversy, as the analogy to chemically based synthetic drugs was much clearer (48). However, new questions have been raised for disease gene patents and "upstream" patents on research tools (any DNA sequence that has a use in research) and broadly defined DNA-based diagnostic tests.

DNA patents granted for research tools and diagnostic uses have become increasingly common, with the majority of these filings occurring in the U.S. (49). Legal experts have argued whether these categories of patents meet the required criteria of novelty, inventiveness, and utility. Putting aside the belief that the USPTO, in particular, has set the bar too low on these standards, convincing arguments have been made that the current legal criteria have been met (50,51). Whether the patent system is operating ethically is a different question, and one that can be assessed through a balancing test where the benefits to the patent holder are weighed against the benefits to society. In the case of patents on therapeutic proteins and their corresponding DNA sequences, the benefits to society in terms of new medicines for serious diseases appear worth paying higher prices to the patent holder. Much less clear-cut is the patenting of research tools where thousands of patents have been granted for sequences easily generated with high-throughput sequencing techniques with weakly demonstrated or speculative uses (52). The scientific community has argued that greater progress will be made if these tools remain freely accessible, allowing different researchers to explore different potential solutions to important medical problems. This view has been cautiously supported by the biopharmaceutical industry through the formation of the SNP Consortium, where the overall intellectual property objective is to maximize the number of SNPs entering the public domain at the earliest possible time (53).

As patent protection moves "upstream" from products to gene–disease associations to research tools, there is the increased potential for stifling innovation and inhibiting research because of the hurdles associated with royalty stacking (54). Whenever there are multiple owners (as is often the case in gene patents), there is the creation of the "tragedy of the anticommons"—this phrase describes the paradoxical situation where awarding patents to multiple owners leads to an underutilization of scarce resources because the owners can block each other (55). In the case of gene patents covering diagnostic testing, there is already evidence that this type of patent is limiting research (56). Another argument against disease gene patents focuses on the inevitable conflicts that will occur for clinicians ordering genetic tests in clinical practice. Because genetic test results nearly always give important, usable clinical information on risks and traits other than those being tested for, physicians will be forced to choose between patent infringement and malpractice. When faced with the need to pay royalties for all patents applicable to a given gene locus, the cost of a given test becomes prohibitively expensive. The clinician's dilemma results from the strict enforcement of disease gene patents and creates the perverse situation where the more we know about a genetic test, the less usable it becomes (57).

The Nuffield Council on Bioethics issued a comprehensive report in 2002 on the ethics of patenting DNA (58). This group also concluded that the potential harms to the public interest through the liberal granting of patents were greatest for gene patents pertaining to research tools and diagnostic tests (in that order), although more stringent criteria should be applied to all future gene patent applications. In particular, the council suggests that the threshold for meeting even the revised *Utility Examination Guidelines* of the USPTO (58a) is still set too low, and many patents have been granted where "credible utility" simply means a "theoretical possibility." Others have argued that the patent system is working well and without more evidence of real, not just hypothetical harms, there is no ethical justification for changing the system (59).

10.3.2 Recommendations

Given the legal costs associated with patent challenges, as well as the strong political and economic interests advancing the current patent agenda, radical changes in patent law are unlikely (60). We have also made the case that radical changes are unwarranted, as there are significant benefits to society for granting gene patents. The ethical gray zone emerges where the evidence of benefit becomes more debatable, for example, as patents move away from products and toward platform technologies. The question then becomes how best to address the ethical concerns associated with gene patents. While much of the attention has focused on intellectual property law and the practices of patent offices in the U.S., Europe, and Japan, in reality neither the law nor patent officials are structured to handle ethical issues (61). Indeed, at least for the USPTO, there is no statutory authority to consider these matters of optimal social policy (62). Any significant changes to the current approach to gene patenting will only be achieved through Congress in the form of new legislation governing biotechnology-based research.

Nevertheless, a number of principles have been identified to guide patent law reform. These include using more stringent criteria for determining inventiveness, novelty, and utility and mandating broader licensing policies to avoid the harms associated with overly broad patents and their dampening effects on research (63,64). In parallel, more studies need to be conducted to measure the real benefits and harms associated with gene patents, so that we can move from emotional arguments to evidence-based discourse (65). Finally, we should continue to promote dialogue and debate regarding gene patents from a wide range of stakeholders.

10.4 Research Ethics

One of the hallmarks of research in biotechnology is the high level of financial and scientific interdependency between industry and university researchers. However, the strong economic ties between academia and the biopharmaceutical industry are often viewed as a double-edged sword. Although these collaborations are typically pursued for their synergistic effects in advancing the progress of biomedical research, increasing levels of industry funding for university-based investigators have simultaneously raised concerns about conflicts of interest and threats to scientific integrity. In the U.S., the rapid growth of these relationships is the result of explicit federal policies and their track record of success in producing health and economic benefits (66).

Historically, university researchers in the life sciences have enjoyed rich federal funding and the freedom to pursue basic research in biological mechanisms underlying health and disease without having to prove the short-term applicability of their work (67). Concerns about the limited translation of federally funded research into useful health care interventions led to the passage of the Bayh Dole Act in 1980, which allowed universities and individual scientists to patent inventions made with federal support, to license these patents to companies, and gave the federal government the right to intervene if those licensees failed to commercialize the resulting products (68). The rationale behind this new legislation was both to foster and reward translational research and new product introductions, and to lessen universities' dependence on federal grants. Additional incentives to patent the findings of publicly funded research were provided by the Federal Technology Transfer Act of 1986 (69), which encouraged commercialization of the results of research conducted by federal laboratories such as the National Institutes of Health (NIH). In parallel, a variety of legal, political, and economic factors enabled greater collaboration between universities and industry, including the 1980 Supreme Court decision to allow the patenting of new life-forms created by biotechnological techniques (70).

As a result, there have been major increases in university patenting and licensing activities (71), and financial relationships among industry, investigators, and academic institutions have become commonplace (72). The growth of these relationships in the U.S. can be measured in terms of dollars (73), the number of strategic alliances (74), the number of patents and licensing agreements (75), and participation by university faculty and their parent institutions in the founding of new start-up companies (76). Indeed, for the biotechnology industry specifically, the practice of scientists starting a company in academia and maintaining equity ownership in the spin-off company while remaining on faculty is the norm (77). The financial situation for smaller biotechnology companies is even more complicated, as a major, or even the dominant, funding source is venture capital (78), with the attendant financial expectations and pressures related to investment bankers. While private investment in university-based biomedical research is a more well-established phenomenon in the U.S. than in Europe (79), similar trends are occurring abroad and are inevitable given the global nature of the biopharmaceutical industry.

10.4.1 *Benefits of Academic–Industry Collaborations*

Industry pursues academic collaborations primarily for access to new ideas, specialized talent, cutting-edge knowledge, and aid in recruiting talented researchers (80). The form of the relationships varies considerably, including research collaborations, consulting, faculty equity, and training (81). For smaller companies in particular, industry expects to license products with immediate commercial value, although most companies have realized patents, products, and sales as a result of these relationships (82).

Universities want access to research funding, particularly as competition for traditional public sources of support has intensified. In both the U.S. and Europe, governments are encouraging academic institutions to partner with industry (83), and there are societal expectations that universities become an engine of economic development (84). More distinct to the field of biotechnology, university-based researchers frequently take equity and leadership positions in start-up companies while maintaining their faculty appointments (85). Since the researcher's parent institution typically owns the intellectual property on which these new companies are based, the university also takes an equity stake in the new venture, arguing that it is entitled to share in the financial upside (86).

Another reason for academic–industry collaborations is the growing scale of research that requires the complementary skills and resources of both parties for studies to be carried out successfully (87). For example, research in genomics requires access to sophisticated techniques and complex equipment, while correlations of genotype with phenotype require access to large, well-characterized clinical populations (88). For these relationships to work, typically the flow of information is bidirectional (89), with industry being recognized for bringing more to the table than just money.

The immediate benefits of these relationships include economic support for universities, researchers, and graduate students, in terms of both study grants and training opportunities (90). Universities also realize royalties on patents and licenses, although this revenue stream is tiny compared to their overall research budgets (91). Academic–industry collaborations may also translate into job creation and economic development both regionally and nationally. For example, some state governments are pressuring or providing incentives for universities to form collaborations with industry, such as Florida's effort to recruit Scripps to open a campus in that state. In the longer term, there is improved transfer of medical innovation to clinical practice (92), with the associated sales revenue for marketed products accruing primarily (but not exclusively) to the industry partner.

The accelerated knowledge generation and translation into products that are facilitated by these partnerships allow policy makers to increase the public health and economic return on their investment in academic research (93). In the case of shared assets, there is the creation of a critical mass to conduct research and development (94). Biopharmaceutical companies contribute important nonproprietary research findings at no cost to the taxpayer (95). Even in the context of proprietary research, private equity firms' contribution to mapping the human genome led to completion of the map years ahead of schedule (96). Although often overlooked, there are also the intangible benefits associated with academic and industry researchers gaining insights into their respective cultures and being willing to learn from each other (97).

10.4.2 Ethical Issues

Despite the broad range of potential benefits of closer financial ties between industry and academia, there are widespread concerns about the possibilities for a variety of harms due to conflicts of interest (98–100). Conflicts of interest are defined as "a set of conditions in which professional judgment concerning a primary interest (such as a patient's welfare or the validity of research) tends to be unduly influenced by a secondary interest (such as financial gain)" (101). It is not necessary that there actually be a negative outcome; a conflict occurs whenever primary and secondary interests coexist (102). Conflicts of interest are ubiquitous in biomedical research and cannot be avoided (103). Many of the examples of nonfinancial conflicts of interest fall under the category of academic self-interest, including desire for faculty advancement (tenure), publication, and peer recognition. Nevertheless, many experts have attempted to make the case that financial conflicts of interest are particularly pernicious and require targeted management. A frequent argument is that financial conflicts of interest are nonobligatory and often unrecognized unless disclosed (104).

Another rationale is that the opportunities for academic investigators to realize large financial gains relatively easily are unprecedented, and these high personal financial stakes have reached a level where the current safeguards mitigating conflicts of interest are viewed as inadequate (105). Finally, others argue that academia has much more experience with handling nonfinancial conflicts of interest, whereas novel approaches are required for financial conflicts of interest because so much private money is flowing into academia so quickly that the former equilibrium is being disturbed in potentially deleterious ways (106). Perhaps the most compelling argument focuses less on trying to rank the negative effects of the various types of conflict of interest and more on the practical reality that "the oversight of nonfinancial conflicts traditionally has been left to the academic community and the professions, but during the past decade financial conflicts have become a shared and contingent responsibility of academe and the federal government" (107). For better or worse, financial conflicts of interest are in the spotlight.

Despite protestations to the contrary, there is compelling evidence that industry funding has a differential impact on both investigator behavior and published research outcomes. Investigators with industry support are more than twice as likely to engage in activities to ensure trade secrecy, such as withholding results from colleagues as investigators without industry support (108). While the need to keep information confidential for more than 6 months in order to file a patent is a common explanation for publication delays (109), this time frame is often exceeded in industry-sponsored research, and extends beyond the National Institutes of Health (NIH) recommendations of a 30- to 60-day period to delay the release of information. Although certainly not their intention, whenever scientists withhold useful information, there is the potential to slow research and prolong human suffering. Secrecy requirements of industry also have a negative impact on graduate students and postdoctorates, because prompt publication of research results can impact their employability after completing their training (110). Few conflict-of-interest policies specifically address the situation where faculty advisors have students working on industry-funded projects, a particularly striking omission given the central role of education and training in the university mission (111).

Pharmaceutical industry spending on research and development currently exceeds the total NIH operating budget ($32 billion vs. $24 billion in 2002), although only a small part of NIH's total budget is focused on pharmaceutical research (112). The majority of industry dollars channeled through universities is for clinical research and in particular clinical trials. Numerous studies have been undertaken to examine the relationship between pharmaceutical industry sponsorship of clinical trials and bias. A recent review paper on this topic revealed that studies funded by the pharmaceutical industry were more likely to have outcomes that favored the sponsor's product than research funded by other sources (113). This was not due to differences in the methodological quality of the studies, but may be explained by inappropriate choice of comparators or by publication bias (114). Failure to report the results of negative studies (either no difference between treatments or the sponsor's drug was less effective) is viewed as unethical because it both violates the implied contracts with patients who participate in these trials and leads to biased estimates of treatment effects (115). A related ethical concern is that many industry-sponsored trials violate the principle of equipoise and thereby threaten the entire clinical trial enterprise (116). (Equipoise refers to the requirement for substantial uncertainty in the scientific community about which of the trial treatments would benefit patients and forms the ethical and scientific foundation for the conduct of randomized trials.)

Industry funding of academic biomedical researchers also threatens to alter the norms and values of universities, which can lead to a slowing of scientific progress and erosion of public support for universities and their freedom to determine their routes of scientific inquiry (117). As evidence of this trend, the research agenda of universities has shifted, with a greater emphasis on clinical research over basic research (118,119) and priorities driven by the potential for commercial applications (120). Market-driven research agendas are an even greater problem for researchers studying diseases prevalent in less developed countries, as it is unlikely that their work will be funded by industry (121). As the lines between academia and industry become blurred, observers fear that there is already a dangerous erosion of the academic values of communalism, pursuit of knowledge unencumbered by the profit motive, and scientific integrity (122).

Academic medical centers are also public resources, entrusted to treat patients, train health professionals, and conduct research ethically. Concerns about the insidious influence of industry funding on all three of these activities are well described (123). Perhaps more compelling to the public are the actual cases of scientific misconduct and investigator bias, particularly those involving actual harm to patient subjects (124). The impression being created is that the academic biomedical enterprise is primarily self-interested, rather than focused on meeting the unmet medical needs of patients. This perception has led to an erosion of public trust, which, unless addressed, is likely to translate into a lack of public support and respect for clinical research (125).

Although certainly not the only source of conflicts of interest, industry funding of university-based research leads to both individual and institutional conflicts of interest that must be recognized and managed to avoid the loss of public trust. However, the social contract between universities and the public also requires that universities are accountable for the taxpayer dollars they receive—they fulfill this requirement by actively engaging the practical application of their work, which in the U.S. most often translates to cooperation with industry (126). Any attempts to foster research collaborations between academia and industry must recognize the deep differences in missions, cultures, and incentives that exist between the two and ensure that the ability of universities to fulfill their role as a public resource is not compromised (127).

10.4.3 Current Safeguards

Over the past 30 years, the responses of the scientific community to the problems of financial conflicts of interest have cycled through two opposing approaches: the prohibition model and the disclosure and peer review model (128). Guidelines proposed in 1989 by the NIH followed the prohibition model and were subsequently withdrawn because they were viewed as unnecessarily restrictive (129). Since adoption of the Public Health Service and National Science Foundation guidelines based on the disclosure model in 1995 (130), the emphasis has been on recognizing and managing both individual (131) and institutional (132) conflicts of interest. The typical management strategy includes requirements for disclosure, a mechanism for auditing adherence to these policies, procedures for granting exceptions, and sanctions for noncompliance (133). As most of these strategies are discretionary, there is tremendous heterogeneity in both the content and application of these policies (134). General principles have been proposed to guide policy development, including the need for formal oversight mechanisms (135). Institutional Review Boards (IRBs) are one potential oversight group; however, policing commercial relationships is beyond their original mandate and they are often viewed as less effective than stand-alone conflict-of-interest committees (136).

Institutional conflicts of interest are particularly problematic and often require formal separation methods—for example, creating independent review panels that oversee the separation of research activities from institutional investments (137). Institutional decision makers and IRB members have a heightened responsibility to divest their financial ties or recuse themselves from responsibility for research oversight given the scope of their responsibilities (138). Peer review represents another check in the system, and journal editors have also developed financial disclosure polices for authors submitting and publishing manuscripts (139,140). Concerns about the harm from conflicts of interest to human subjects participating in research have led to calls for strengthening the current research protections (141), and a proliferation of conflict-of-interest guidelines pertaining to clinical research (142). As the likelihood of harm to patients decreases for translational and basic research, more liberal approaches have been recommended for nonclinical research (143). The goal is to develop strategies for managing financial conflicts without unduly impeding the creative social benefits of academic–industry collaborations (144).

10.4.4 Problems with Current Management Strategies

There is no shortage of innovative, flexible, and practical recommendations for how to manage all types of financial conflicts of interest (145–151). Leaders in the field have moved well beyond descriptive studies of the problem and hand-wringing to real solutions. The most difficult step is in figuring out how to apply the principles and policies to individual cases in a consistent and transparent fashion. Evidence abounds for uneven adherence to prevailing policies at various levels of the research enterprise up through and including publication of research results in peer review journals (152–157). Although IRBs are frequently expected to include evaluation of financial conflicts of interest as part of their assessment of risks and informed consent, these groups may lack the expertise to fulfill this role and are already viewed as overburdened (158) and underperforming (159).

The heavy reliance on the practice of disclosure (to university officials, to journal editors, and to readers of peer review publications) can give participants a false sense of security by emphasizing financial conflicts of interest while overlooking other equally or even more important conflicts of interest. Indeed, the very notion that disclosure neutralizes bias has been challenged as a fallacious concept (160). In addition, very little is known about how individuals interpret and process conflict-of-interest disclosure information (161). It is important to recognize that "disclosing a conflict only reveals a problem, without providing any guidance for resolving it" (162). Focusing on financial disclosures may actually cause harm, by diverting the reader's attention away from the rigorous scrutiny of scientific methods and oversimplifying a complex issue (163). The other unintended consequence of censoring the input from individuals with industry affiliations is that objectivity is compromised by limiting free and open scientific dialogue (164).

10.4.5 Recommendations

The "corporatization" of biomedical research is a trend that is unlikely to be reversed given the current incentive structure in the U.S. and public expectations for new and better medical treatments. However, it is equally unlikely that researchers and clinicians can wear two hats (inventor/entrepreneur and researcher/clinician) and maintain the trust of patients and the public (165). Despite the limitations of the disclosure and peer review model for managing academic–industry financial relationships, it is the most feasible and ethical approach because it attempts to balance the benefits and risks of these relationships.

However, simply adding a layer (or multiple layers) of oversight to the current medical research enterprise is an insufficient step. At a minimum, there needs to be greater consistency in conflict-of-interest rules across institutions (166). An example would be the 2001 Association of American Medical Colleges (AAMC) guidelines, but these only cover individual financial interests in human subjects' research—standard approaches to institutional conflicts of interest and conflicts of interest in translational and basic research need to be developed (167). More attention needs to be focused on how these guidelines are implemented and appropriate requirements built into the contracts before they are signed by industry and university representatives. A model for this approach is the practices followed by the Howard Hughes Medical Institute, which strives to avoid relying on a "bolting the barn door after the horse has fled" disclosure strategy (168).

These policies need to be complemented by consistent and transparent disclosure guidelines for peer review publications (169). To avoid the related concerns about publication bias and variable clinical trial quality, mandatory clinical trial registries (170) should be seriously considered and all trials results reported in a standardized fashion (171).

To increase the likelihood that these oversight practices will be voluntarily followed in the absence of federal regulation, oversight committees such as IRBs will need to receive additional training and support, including financial support. There is already the perception that worthwhile studies are not pursued because the burden of current regulatory compliance has become too high (172). Adding management of conflicts of interest to IRBs without additional resources is a formula for superficial and inconsistent compliance. While these measures will obviously add to the costs of conducting research, relative to the billions of dollars spent on industry-sponsored research, the incremental costs are small and the potential savings in terms of harm avoided are huge.

The attention placed on institutions and organizations should not overshadow the need to focus on the ethical responsibilities of individuals, particularly the scientists and clinicians that are leading the projects. They need to view themselves as accountable for both the technical and ethical integrity of their studies. The ability to identify potential ethical dilemmas and participate in the development and implementation of corresponding solutions should be built explicitly into the job description of researchers, regardless of their work setting. These efforts should be supported by universities in their training programs and by professional societies.

Research is needed to understand the effectiveness of disclosure practices. For example, how do patients, university decision makers, investigators, peer reviewers, editors, and the public (including the media) interpret and apply information about financial conflicts of interest? Do their responses differ when provided with information about other types of conflicts of interest? If disclosure is to be the primary safeguard in the current system, then it too must be studied before receiving broad approval.

None of these attempts to manage conflicts of interest will have the desired effect without extensive educational efforts focused not just on researchers in training (173), but also on industry employees and their investors, the press, legislators, regulators, and consumers (174). As all of these groups have competing priorities and needs, simple solutions will not suffice and difficult choices will need to be made (175). Over time, the cultural divide that separates academia from industry will become smaller both through education and from cross-fertilization of ideas from individuals who work in multiple sectors throughout their professional careers.

The reality (at least in the U.S.) is that the proportion of biomedical research funded by industry will only get larger in the future (176). Most NIH funding is for basic research (177), and there is little chance that expensive drug clinical trials will be funded by sources other than the sponsors that have a vested interest in the results (178). As the evidence-based medicine movement gathers steam, the demand for large (and expensive) clinical trials will only grow. Given the tremendous productivity of the biopharmaceutical industry (179) and university researchers (180), much in the current system appears to be working. Nevertheless, major challenges exist for the national clinical research enterprise, including lack of funding, regulatory burdens, fragmented infrastructure, high costs, slow results, and a shortage of qualified investigators and willing participants (181). These obstacles are impeding the translation both of basic science to clinical studies and of clinical studies into medical practice. Solutions will only come from the collaborative efforts of the various stakeholders in the system—proper attention to conflicts of interest will facilitate these collaborations.

10.5 Genetic Privacy in Clinical Practice

The enthusiasm surrounding the potential for genetic advances to improve human health is often tempered by public concerns about privacy infringements and discrimination based on genetic information (182). These concerns stem from the public perception that DNA possesses greater power and significance than other medical information, a notion that is often reinforced by how scientists and the media interpret progress in genetic research (183). Unlike an individual's medical record, which contains information about his current health and medical history, DNA has been characterized as a future diary (184) that contains risk information about diseases that are likely to occur in the future. As DNA residing in stored blood or tissue samples can be tested at any time, the DNA molecule itself might be viewed as a medical record, containing unique predictive information that could be used by insurance companies and employers to discriminate against currently healthy individuals (185). Without a clear national policy about the collection, use, storage, and protection of genetic information, many people in the U.S. fear that their DNA will be tested against their will or without their knowledge, or that genetic test results will be misused (186). These unauthorized or inappropriate uses of genetic testing are viewed as significant threats to individuals' privacy rights (187).

10.5.1 *Ethical Issues*

Privacy is a complex notion that has been characterized along four dimensions: informational, physical, decisional, and proprietary privacy (188). Informational privacy describes the individual's ability to control third-party access to his personal health information and is what most people think of when they describe the right to genetic privacy. Physical privacy refers to restricted access to persons and personal spaces. Concern about interference with personal choices from third parties such as the government is covered by the concept of decisional privacy, and control of ownership of personal materials and information is the focus of proprietary privacy (189). A term that is often used interchangeably with privacy is confidentiality; however, they are distinct concepts. Confidentiality refers to the right of an individual to prevent the redisclosure of private information that was shared within the confines of a professional, fiduciary, or contractual relationship (190). With respect to genetic information, the duty to maintain confidentiality typically occurs in the setting of the patient–physician relationship, and much of the debate has centered on how to protect the unauthorized disclosure of genetic test results obtained in the clinical practice setting (191).

Another separate but interrelated concept focuses on how genetic information will be used by third parties, and policies directed at preventing misuse are covered by genetic antidiscrimination laws (192). In the U.S., there currently is a heterogeneous set of state laws and only one federal law that incompletely address various aspects of genetic privacy, confidentiality, and nondiscrimination. The lack of a comprehensive and coherent federal privacy policy that protects all these interests is cited as a major reason why many individuals say they do not want their employer or health insurer to have information about their genetic profile and are unwilling to participate in genetic testing (193,194).

Because respect for privacy is essential for preserving individual autonomy (intrinsic value) and required for both individual and public health reasons (extrinsic value) (195), some experts have advocated for separate laws governing genetic privacy (196–199). The rationale for this position includes (1) the predictive nature of genetic information, (2) the potential for family risks and group harm, (3) the permanency of DNA samples and breadth of information contained within, and (4) concerns about genetic discrimination, particularly given the history of eugenics in the U.S. (200). These arguments form the basis for claims that genetic information represents a distinct type of medical information that merits special protections, a concept often referred to as genetic exceptionalism (201). Others have argued that genetic exceptionalism is not warranted, as it is not possible to clearly differentiate genetic information from other types of medical information (202). They dismiss the notions that genetic information defines who we are as individuals (genetic essentialism) and represents information that is uniquely predictive of our health status throughout our life span (genetic determinism) as overly simplistic and misleading (203). By advocating for special genetic privacy protections, the deterministic nature of genetic information is underscored in a way that is not supported by the science and does more harm than good (204,205). While genetic testing does not create novel ethical dilemmas for clinicians, they are obligated to consider the implications of predictive genetic testing for patients, the net benefits and harms of specific genetic test results, and counsel patients accordingly (206).

10.5.2 Legal Response

In the face of these divergent positions, a patchwork of laws at the state and federal levels have been passed in order to attempt to reassure the public that the privacy of genetic samples and the information obtained from analysis of the genetic material will be protected. Currently, the only federal legislation covering the use of genetic information is in the context of the Health Insurance Portability and Accountability Act (HIPAA), which prohibits health insurance discrimination based on any "health status-related factor" (including genetic information) (207). However, this provision only applies to group health plans (typically covering > 50 members) and there are no protections for the small group or individual insurance markets. The HIPAA privacy regulation is focused on limiting disclosures of individual protected health information (PHI) and does not distinguish genetic information from other types of PHI. However, the regulation does not directly reach all individuals or entities that have access to PHI, such as employers, pharmaceutical companies, and researchers, and therefore has been criticized as offering incomplete privacy and nondiscrimination protections for all PHI, including genetic information.

At the state level, there is a broad range of genetic antidiscrimination legislation; however, even when states such as Massachusetts have passed a comprehensive law including both privacy and antidiscrimination protections, closer analysis reveals significant weaknesses and gaps (208). Examples of these deficiencies include failure to protect storage and retesting of identifiable

DNA samples and lack of strong penalties for privacy infringements (209). The majority of states strictly prohibit the use of genetic information for risk selection and risk classification purposes for health insurance. However, the Employee Retirement Income Security Act (ERISA) permits self-funded employee benefit plans to preempt state insurance regulation, so employees of large employers are not protected (210). Laws in 16 states require informed consent before a genetic test can be performed, and 17 states have established either civil or criminal penalties for violating genetic privacy laws (211). With respect to use of genetic information by employers, 31 states have some type of genetic nondiscrimination in employment laws, but the scope and functions of these laws vary greatly (212). These state laws have been criticized for (1) their presumption of genetic exceptionalism, (2) their inability to overcome the inherent difficulty in distinguishing genetic information from other types of medical information, (3) their potential for creating adverse selection for the insurance industry, and (4) their lack of fairness in preventing discrimination against those with diseases that (currently) have no known genetic basis (213). For example, these laws would protect individuals who tested positive for a genetic predisposition to colon cancer, but not individuals with a precancerous polyp with no known genetic origin. The concern that people should not be responsible for inheriting certain genetically based conditions reinforces the erroneous assumptions of genetic determinism and essentialism while leaving unchallenged the converse inference that people are responsible for diseases that occur due to chance or environmental conditions.

Perhaps most remarkably, this proliferation of genetic privacy and antidiscrimination laws has occurred in the absence of any compelling evidence that health or life insurers are using genetic information for underwriting purposes (214). This is because genetic tests results for presymptomatic diseases are not viewed by insurers as economically important or useful, particularly given the short time horizon used by underwriters for predicting costs and high turnover rates in the insured population (215). Moreover, there apparently is no correlation between variation in state genetic antidiscrimination laws and the occurrence of insurance discrimination across states (216). This is not surprising, given that the actual rate of insurance discrimination is very low, despite earlier claims that genetic discrimination was extensive. This discrepancy most likely resulted from serious methodological flaws in many of the previous studies (217,218). What is equally problematic is that there is no evidence that additional or different legal protections are likely to increase patient willingness to undergo testing (219). Thus, it appears that policy makers have proposed solutions to an essentially theoretical problem, perhaps based on their lack of understanding of genetic information or because of overstatements by genetic experts about the predictive power of genetic test results (220).

With respect to the workplace, similar concerns exist that employers will misuse genetic information to discriminate against employees (221). Although there are a few well-publicized examples, such as the case of the Burlington Northern and Santa Fe Railway conducting genetic testing of employees with carpal tunnel syndrome without their knowledge (222), there is no evidence of widespread use of genetic testing by employers (223). While the basis for the legal challenge was the Americans with Disabilities Act (ADA), the case was eventually settled out of court, and it remains unclear whether the courts will determine whether the ADA should be interpreted to prohibit employee discrimination on the basis of a presymptomatic genetic condition. Although the ADA is frequently cited as potentially providing safeguards against employer use of genetic information to make hiring, promotion, and discharge decisions, recent erosions in the courts' interpretation of a disability makes this a weak protection at best (224).

10.5.3 Problems with Current Legal Remedies

Despite their intuitive appeal, laws requiring special treatment of genetic information are harmful on several levels:

- They reinforce beliefs in genetic determinism and ascribe an inflated level of importance to genetic risks that is not supported by the science.
- They reinforce concerns about employment and insurance discrimination in the absence of any real evidence of current or past misuse of genetic information for underwriting or hiring decisions (225).
- They interfere with the quality and efficiency of care because of the isolation and differential treatment of genetic information from the patient's central medical record.
- They lead to increased direct and indirect medical costs associated with enforcing separate genetic privacy regulations.

Perhaps the most positive view of genetic antidiscrimination policy efforts to date is that they have codified social norms regarding prohibitions against the future use of genetic information by insurers and employers (226). Having these laws on the books reflects current public attitudes and establishes the social illegitimacy of using presymptomatic genetic information by insurers and employers. A related positive interpretation is that if we wait until there is real evidence that genetic discrimination is occurring, it will be too late (227). Better to be proactive, particularly in an era of computer databases and relatively easy electronic access to medical information, and establish strong privacy and antidiscrimination protections before real problems occur.

Our position is that the ethically justifiable position (and a better use of scarce health care resources) is to advocate for ensuring the privacy, confidentiality, security, and appropriate use of *all* types of personally identifiable medical information, rather than carving out separate protections for genetic information. The major ethical concern is the misuse of genetic information, and this is best addressed through antidiscrimination legislation. However, this legislation is only appropriate if it treats all types of medical information in a similar fashion and does not single out genetic information for special treatment. Although the debate continues, what is clear is the need for additional genetics education across all stakeholder groups. Only on the basis of the most up-to-date understanding of the contribution of genetic risk to human health and disease will we be able to strike the appropriate balance between ensuring the privacy of medical information and its fair use.

10.6 Ethical Issues in Genetic Research

Demonstrating respect for persons and avoiding harm to them are well-established ethical principles in contemporary clinical research and are ensured through the process of IRB review and informed consent. As one part of their review, IRB members have traditionally focused on weighing whether the research-related harms to the individual outweigh the potential benefits of the study to society. Up until the early 1990s, IRBs have assessed potential study-related harms primarily (although not exclusively) in terms of physical harms, such as threats to patient safety (228). With advances in genomic medicine, the evaluation of study subject harms has been expanded to include informational harms that may result from the inappropriate disclosure or use of genetic information (229). Although historically the harm associated with genetic information obtained in the research setting has received relatively less attention than genetic information acquired in the clinical setting, within the past decade this issue has become a major bioethical concern. Concerns about the lack of assurance of the privacy of genetic information obtained from research participants, the possibility of subsequent employment and insurance discrimination, and stigmatization of racial or ethnic groups who participate in these studies now dominate the ethical evaluation of genetic research. In particular, the widespread availability of human tissue for genetic analysis, the ease with which genetic information can be shared electronically, and the commercial potential of genetic information linked to clinical characteristics have raised concerns that the current system of research protections is inadequate for genetics research (230).

An extensive review of the federal regulations governing human subject research is beyond the scope of this chapter. In addition to the Common Rule (231) (the federal regulation that governs research on human subjects by institutions that receive federal funding and that has established a *de facto* standard in the U.S. even for nonfederally funded research), the FDA plays a major role in regulating human subjects' participation in clinical trials involving investigational new drugs (232). Protocols for the collection and analysis of genetic samples as part of pharmaceutical clinical trials are becoming routine within the industry, particularly as the science of pharmacogenetics improves the drug development process. There appears to be an emerging ethical standard within the biopharmaceutical and private DNA banking industry that includes IRB review, separate informed consent forms for participation in the genetic analysis component of the clinical trial, and stringent privacy safeguards, such as the use of firewalls and anonymizing data (233).

Much greater variability in the conduct of genetic studies on stored tissue samples exists in academic settings, where arguments have centered on striking the appropriate balance between protecting patient privacy and the scientific value of maintaining linkages to relevant clinical information (234). While informed consent and IRB review are the unquestioned cornerstone of research involving human subjects, the use of stored tissue samples for genetic studies represents an ethical gray zone where patients are frequently not aware of how their tissue samples will be used for genetic research and have not given their explicit consent for these experimental applications.

Several groups have been convened at the federal level to address this complex topic, with an influential series of recommendations emerging from these expert panels despite the absence of formal legal authority (235,236). These recommendations attempt to balance the interests of the research community, which advocates the benefits of genetic research using stored tissue samples, with the interests of individuals and groups that may be harmed by the results of this research. Generally, they raise the bar for the use of informed consent and IRB oversight and discourage the routine use of IRB waivers. More recently, guidance has emerged on the use of informed consent for population-based observational genetic studies (237). The same ethical concerns highlighted in clinical trials apply to epidemiologic studies—obtaining informed consent for DNA banking, protecting privacy and confidentiality, and avoiding family or group harms (238). Current best-practice recommendations for genetic research include creation of a one-way link for human tissue biobanks (239), informed consent for coded, linked, or identifiable patient samples (240), and privacy protections for research data that exceed the current standards for medical records (241).

Given the growth of commercial interests in genetics research and the use of indigenous populations for genetic studies, there is a series of related ethical concerns that expand beyond considerations of autonomy to include considerations of justice. For example, controversy has surrounded both government-sponsored and commercially sponsored projects to create databases of genetic variation linked to phenotypic information for defined populations in Great Britain (242), Estonia (243), and Iceland (244). Beyond the debate over whether participating citizens have been able to give adequate informed consent (245), there are additional concerns about commercial interests diverting attention from more immediate national health problems, such as tobacco abuse, whether there will be a sufficient return of research profits back to the participating countries, and the possibility for stigmatization based on research findings. These novel research approaches have expanded the bioethical discourse to include discussion of the potential for economic and psychosocial harms (246).

To address these issues, experts have called for explicitly recognizing the contributions of individuals and groups participating in genetics research (247), as well as developing greater sensitivity to community interests and greater involvement of community leaders in the research endeavor (248).

For example, results demonstrating a correlation between a specific disease susceptibility and a particular racial or ethnic group could lead to stigmatization and discrimination. The need to review study protocols with leaders of the community and gain their approval before embarking on a genetic study has been challenged on the grounds that it is difficult to know who is qualified to speak for the community of interest and it is impossible to obtain group consent to a study (249). Others have suggested that the impracticalities of operationalizing community review of research proposals can be avoided by adopting a community consultation model where the degree of interaction with the community is tailored to the risks of the particular study (250). In this model, the benefits of interacting with the community during the planning phases of a study, in terms of gaining their support, purportedly outweigh the harms to the study design, in terms of escalating costs and time delays (251).

10.6.1 Recommendations

Genetic research has the potential to greatly improve our ability to understand and treat human disease. Society has traditionally valued biomedical research and the altruistic contributions of those who participate in research as subjects. However, without adequate attention to issues of autonomy, justice, and nonmaleficence, trust in the research enterprise will be undermined and the social goals of research will not be realized. Research in genetics is particularly emotionally charged for the reasons described above and represents a compelling case for the application of proactive bioethics to avoid damage to the field because of insensitivity to social concerns and cultural values (252). To ensure the trust and continued involvement of research participants, we recommend the routine use of informed consent that includes general information about all potential disclosures and the nature and magnitude of risks associated with these disclosures (253). There should be no blanket-use authorizations included in informed consents (254), and IRBs should pay particular attention to parental authorization procedures in studies that plan to conduct genetic testing of minors. Researchers should not enter individually identifiable genetic research results not utilized for health care in the medical record (255). Researchers should also develop realistic procedures that allow study participants to withdraw their consent once the study has started. Serious consideration should be given to the use of third-party brokers of genetic information who would be responsible for obtaining DNA samples, medical histories, and informed consents from individuals and then making this information available to researchers in a coded form (256). Individual study protocols that access these genetic data banks would then be reviewed and approved by IRBs (257). Finally, there should be adequate measures in place to ensure compliance with these practices and remedies for individuals whose rights are violated (258,259). With proper attention to these ethical safeguards, researchers will be able to balance pursuing the benefits of genetic research with public concerns.

10.7 Genomics and Hype

The excitement and media attention surrounding the Human Genome Project are directly attributable to the rosy predictions made by leaders in genetics about the breakthrough medical benefits that will result from our growing understanding of molecular biology (260). Mapping the human genome has led many experts to envisage that genetics will profoundly change the practice of medicine over the next several decades (261). One of the areas of transformational change will be to move Western medicine from "sick care" to predictive medicine, where entire populations of healthy people will be screened for genetic risk factors in order to customize interventions and prevent disease (262). This disruptive innovation has been characterized as an information revolution and one of the greatest scientific revolutions in history (263). Others have suggested that genomics has taken center stage in both basic science and clinical medicine and "is rapidly becoming the primary way in which biology engages with the everyday practical world" (264). Because medicine is an applied science, physician-researchers have been unable to resist the temptation to extrapolate from structural genomics to new cures, despite the lack of a road map to navigate this divide (265).

The pharmaceutical industry is similarly enthusiastic about this new field, particularly as the industry is badly in need of improvements in drug discovery and development and is looking to pharmacogenomics as a critical source of innovation (266). In the face of concerns about gaps and weaknesses in the current pipeline of new drugs, there is optimism that improved understanding of the genetic and molecular basis of disease will result in many new potential drug targets and will reshape the industry's approach to drug discovery (267).

Pharmacogenomics is also predicted to enable much needed improvements in the drug development process. Genetic profiling of potential clinical trial participants to enrich the study sample with likely responders, while deselecting participants at increased risk of adverse drug reactions, will reduce the sample size required to show clinical benefit. Similarly, there will be a greater likelihood of clinical trial success given a genetically homogenous trial population, including potentially the "rescue" of drugs with formerly unacceptable adverse event profiles (268). Overall, the development process will become more efficient as clinical trials will be smaller, faster, and less expensive to conduct (269).

With respect to clinical practice, experts have forecasted that pharmacogenomics will significantly improve how drugs are prescribed in the future (270). Despite major advances in drug therapy for many diseases, treatment remains suboptimal for a significant proportion of individuals because of unpredictable side effects or lack of response (271). Understanding the relationship between SNP maps and drug response will allow clinicians to tailor drug therapy to individual patients and avoid the morbidity and costs associated with adverse drug reactions or lack of effectiveness (272). This issue is particularly relevant given the widespread call to action by clinicians and health care payers in response to the findings that adverse drug reactions occur frequently (273,274), and there are enormous costs associated with managing the downstream consequences of both drug toxicity and lack of response (275). The promise of pharmacogenetics is that clinicians will be able to customize drug therapy to the individual, rather than continue their current practice of empiric prescribing based on published data of the average effect in clinical trial populations, or based on anecdotal experiences (276).

Against this very optimistic backdrop, a few cautionary notes have begun to enter the scientific literature. By extolling the power of DNA for predicting disease risks and developing new drugs, we are reinforcing the misleadingly simplistic notions of genetic determinism and essentialism (see earlier discussions of this topic). For most common chronic diseases, such as hypertension and diabetes, simple relationships between a single gene mutation and the clinical manifestations of the disease do not exist (277). Rather, multiple, incompletely penetrant alleles are most likely involved, and for the population at large, the magnitude of risk conferred by the various genotypes is very low (278). In addition, many of these genetic variants will only increase the risk of disease in the presence of certain environmental or behavioral factors, making the relative contribution of genotype to phenotype very small and unpredictable (279). This makes population-based screening for genetic risk factors impractical because the positive and negative predictive values of the tests will be too low to be clinically useful (280). Finally, there are very few examples of diseases where interventions are available to prevent the disease in currently healthy people with a positive genetic test or to improve their clinical course if the disease eventually developed (281).

Given these current genetic, epidemiologic, and clinical realities, it is not clear that people will want to know their genetic status and elect to be tested (282). Practically speaking, currently there are huge gaps in knowledge about genetics among primary care practitioners (283). In order to fulfill the promise of genomic medicine, these gaps would need to be closed before primary care physicians would be able to appropriately order and interpret genetic tests, as well as counsel patients (284). As an example, a study of the clinical use of commercial Adenomatosis polyposis coli (APC) genetic testing for familial adenomatous polyposis revealed that even when the majority of physicians ordering the tests were specialists (i.e., gastroenterologists and medical geneticists), 80% of the time they did not obtain informed consent or offer genetic counseling before the tests, and the tests results were interpreted correctly only 68% of the time (285). Unless these educational and practice deficiencies are systematically addressed, there is the strong potential that patients will receive misinformation with all the attendant clinical, economic, and psychological harms.

10.7.1 Ethical Issues

The hype surrounding genomics has led to a number of ethical concerns. From a public health perspective, emphasis on the genetic basis of disease has real potential to divert funding and attention away from research and prevention strategies involving known risk factors such as obesity, poverty, and smoking. It is not at all clear that the primary threats to our health are programmed in our DNA, rather than our social environment (286). Emphasizing the deterministic nature of genes in causing human disease is both bad science and bad clinical practice, and may do more harm than good for both public health and distributive justice concerns in the near term.

As described above, numerous problems can emerge from the unmanaged introduction of genetic tests into the marketplace. With a lack of knowledge of both genetics and screening tests, physicians may cause real harm to patients. Some individuals are already predicting that the complexity of genetic information may necessitate the emergence of new medical specialists called "genomicists" or informationalists in order to ensure that genomic medicine will be practiced appropriately (287). Insurers and employers are equally likely to misinterpret genetic test results and may develop coverage or employment practices that are not justified by the science. The situation is only exacerbated by the direct-to-consumer marketing of genetic tests, which inappropriately induces demand for services while potentially providing false positive or false negative results for serious conditions such as breast cancer.

Much of the hype regarding the use of genetic testing to lower unnecessary health care costs has been in the area of pharmacogenomics, particularly given rapidly growing drug expenditures and financial barriers to access to drugs in vulnerable populations such as the elderly. For insurers facing rapidly rising drug costs, one logical application of pharmacogenomic testing will be in a gatekeeper role where drugs will only be covered if the patients are first shown to have a pharmacogenomic test result that indicates the drug is likely to be safe and effective (288). Public or private drug benefit providers may condition coverage decisions on the results of a pharmacogenomic test conducted as part of a precertification requirement (289), again raising ethical issues, as some patients may be unwilling or unable to undergo testing. Mandating pharmacogenetic testing as a condition of receiving coverage for a particular drug could be viewed as a threat to personal autonomy, as well as raise questions about fairness and distributive justice. While it is not reasonable to expect that payers will cover all treatments that might possibly benefit an individual regardless of cost, to date the criteria for limit setting with respect to drug coverage have not been made public to patients. Therefore, pharmacogenetic tests could be perceived by patients as another unjust managed care strategy that denies individuals access to drugs that might benefit them.

With respect to breakthrough research in areas such as gene therapy, hype has the potential to increase the likelihood that researchers and research subjects will fall prey to the therapeutic misconception. The primary aim of research is not to improve the health of research participants, but rather is the search for new knowledge. This distinction explains why research is more heavily regulated than clinical practice and research ethics are modeled on different values and priorities from those governing clinical care. In addition to IRB review of research protocols, the requirement to obtain research participant informed consent is a major strategy to protect individual autonomy and promote rational decision making. Among bioethicists there is agreement that the primary justification for informed consent in research is to enable autonomous choice by research participants.

Despite attempts to clearly distinguish between research and therapy, the line between these two activities has become blurred, particularly with respect to gene therapy research (290). Churchill et al (291) propose an explanation for the conflation: "The cluster of scientific, economic, and cultural hopes swirling around our genes seems to intensify and sustain the future promise of gene therapy at the same time that it frames this revolutionary concept in traditional garb—as merely the next wave of therapeutic options." The harm of attributing individual therapeutic benefit to gene therapy trials is that the patient's autonomy will not be respected, there will not be truly informed consent, and the conflation will result in misrepresentation and disappointment.

10.7.2 Recommendations

While it is too soon to predict whether genomics will represent a disruptive or incremental change to the health care system in the U.S., and the likely timing of this change, there are a number of steps that can be taken in the short run to mitigate potential ethical harms. Clearly, good science (basic, translational, and clinical research) needs to be the foundation for any recommendations, and much work needs to be done to determine the relative contribution of genetics to the risk of disease and the response to drugs. The complex and poorly understood interplay of genetics and environmental and lifestyle factors makes it unlikely that there will be a profound change in clinical practice for at least the next decade, if not longer. This gives us time to make investments to educate all the stakeholder groups about genetics and statistics. Educational programs need to be developed for patients, clinicians, payors, benefit managers, researchers, bioethicists, and investors in the biopharmaceutical industry. Another influential group that requires additional training is the media, who have a large influence on how science is translated to the public.

With respect to cautionary flags about the impact of pharmacogenomics on professional standards of care and coverage decision making, it appears that ethically the trend is in the right direction. For example, everyone would agree that drug-related adverse events are undesirable, potentially life threatening, expensive, and should be eliminated. A related benefit of better medication safety would be improved medication compliance, as intolerable side effects are an important cause of noncompliance. If pharmacogenomic tests are widely available, then it is more ethical to require that patients be tested rather than the current practice of empiric prescribing. Clearly, it is preferable from a moral, as well as economic, point of view *not* to put patients at risk by prescribing drugs that will not work or are unsafe for them. It is important to emphasize that the benefits of this type of application will only be realized if the tests have high positive predictive value; this evidence is not available today, so all stakeholders should be motivated to promote studies of pharmacogenomic test reliability (292). For any type of genetic test, including pharmacogenomic tests, patients should be counseled prior to testing and provided the opportunity to give informed consent.

Regarding genetic therapy research, a number of solutions have been proposed to address the harms of the therapeutic misconception and restore future trials to a solid ethical foundation. These include revisions to the language of informed consent documents to avoid admixing of the aims of research and therapy; descriptions in the informed consent of the evidence for benefit so that participants can make their own assessments of the likelihood of risk or benefit; NIH training programs in research ethics; and creation of an interagency forum to direct conceptual research on the topic and report findings to various stakeholder groups such as the FDA and NIH (293,294).

10.8 Conclusions

While we certainly share in the hope that mapping the human genome will improve our ability to combat disease and improve quality of life, we want to ensure that the associated ethical concerns are explicitly highlighted and potential responses are debated. This chapter has provided a comprehensive overview of the major ethical issues associated with genetics and genomics, beginning with the intellectual property debate and ending with the potential impact of genetics and genomics on the future delivery of health care. Our goal has been to illustrate how ethical considerations are likely to affect not only the application of genetics and genomics to drug discovery, but also drug development and ultimately the clinical use of genetically based drugs.

Advances in understanding the genetic basis of many diseases and the creation of genomic and proteomic databases will contribute significantly to the laudable goal of finding the next generation of novel and more effective drug therapies. However, these breakthrough scientific developments in genomics and proteomics bring with them a host of ethical concerns. While these ethical issues are not unique to the field of human genetics, they have been elevated in visibility and importance because of anxieties that genetic information might be misused or that scientists would be "playing God" and altering the course of human development in potentially harmful ways. To ensure that the benefits of the new genetics can be appropriately realized in drug discovery and development, scientists and senior managers of biopharmaceutical companies must be familiar with the ethical issues and develop strategies to avoid potential problems.

Throughout this chapter, we have attempted to make the case for the value of proactive bioethics for genetics in the areas of both research and clinical practice. By offering a number of practical recommendations for a broad range of situations likely to confront individuals in the biopharmaceutical industry, we believe that steps can be taken to reduce the likelihood of morally problematic consequences while maximizing the social benefits of genetic research.

References

1. International Human Genome Sequencing Consortium. Initial sequencing and analysis of the human genome. *Nature* 409:860–921, 2001.
2. JC Venter, MD Adams, EW Myers, et al. The sequence of the human genome. *Science* 291:1304–1351, 2001.
3. FS Collins, VA McKusick. Implications of the Human Genome Project for medical science. *JAMA* 285:540–544, 2001.
4. RK Dhanda. *Guiding Icarus: Merging Bioethics with Corporate Interests*. New York: John Wiley & Sons, 2002, pp. 1–3.
5. S Gottlieb. The future of medical technology. *New Atlantis* 1:79–87, 2003.
6. RS Eisenberg, RR Nelson. Public vs. proprietary science: a fruitful tension? *Acad Med* 77:1392–1399, 2002.
7. RK Dhanda. *Guiding Icarus: Merging Bioethics with Corporate Interests*. New York: John Wiley & Sons, 2002, pp. 43–47.
8. RK Dhanda. *Guiding Icarus: Merging Bioethics with Corporate Interests*. New York: John Wiley & Sons, 2002, p. 55.
9. For more on these and the biopiracy arguments, see Donald Bruce and Ann Bruce, *Engineering Genesis*, Earthscan, London, 1998.
10. A Gupta. Scientific perception of farmers' innovations in dry regions: barriers to the scientific curiosity, published under the title "Scientists' view of farmers' practices in India: barriers interaction." In *Farmer First*, R Chambers, A Pacey, LA Thrupp, Eds. 1989, pp. 24–30.
11. A Sherma. Tree Focuses Debate on Control of Resources. *LA Times*, 1995.
12. M Blakeney, J Cohen, S Crespi. Intellectual property rights and agricultural biotechnology. In *Managing Agricultural Biotechnology, Addressing Research Program Needs and Policy Implications*, J Cohen, Ed. CABI Publishing and the International Service for National Agricultural Research, The Hague, 1999.
13. P Thompson. *Food Biotechnology Is Ethical Perspective*. London: Blackie Academic, 1997, pp. 163ff.
14. J Merz. Discoveries: are there limits on what may be patented? In *Who Owns Life?* D Magnus, A Caplan, G McGee, Eds. Amherst, NY: Prometheus Books, 2002, pp. 99–116.
15. C Juma. *The Gene Hunters*. Princton, NJ: Princeton University Press, 1989. A Gupta. Rewarding creativity for conserving diversity in third world: can IPR regime serve the needs of contemporary and traditional knowledge experts and communities in third world? In *Strategic Issues of Industrial Property Management in a Globalising Economy*, APPI Forum Series, T Cottier, P Widmer, K Schindler, Eds. Oxford: Hart Publishing, 1999, pp. 119–129.
16. P. Drahos. Indigenous knowledge, intellectual property and biopiracy: is a global bio-collecting society the answer? *Eur Intellect Prop Rev* 22:245–250, 2000.
17. Human Genome Organization (HUGO) Ethics Committee. Statement on Benefit-Sharing. London: Human Genome Organization, 2000.
18. Human Genome Organization (HUGO) Ethics Committee. Statement on Benefit-Sharing. London: Human Genome Organization, 2000.
19. BM Knoppers, on behalf of the Human Genome Organization (HUGO) Ethics Committee. Genetic benefit sharing. *Science* 290:49, 2000.
20. JK Iglehart. Good science and the marketplace for drugs: a conversation with Jean-Pierre Garnier. *Health Aff* 22:119–127, 2003.

21. JF Merz. On the intersection of privacy, consent, commerce and genetics research. In *Populations and Genetics: Legal Socio-Ethical Perspectives*, BM Knoppers, Ed. New York: Kluwer Legal International, forthcoming, 2003.
22. R Chadwick, K Berg. Solidarity and equity: new ethical frameworks for genetic databases. *Nat Rev Genet* 2:318–321, 2002.
23. SM Thomas, MM Hopkins, M Brady. Shares in the human genome: the future of patenting DNA. *Nat Biotechnol* 20:1185–1188, 2002.
24. *Diamond v. Chakarbarty*. 447 US 303 (1980).
25. J Wilson. Patenting organisms: intellectual property law meets biology. In *Who Owns Life*? D Magnus, A Caplan, G McGee, Eds. New York: Prometheus Books, 2002, pp. 25–58.
26. J Wilson. Patenting organisms: intellectual property law meets biology. In *Who Owns Life*? D Magnus, A Caplan, G McGee, Eds. New York: Prometheus Books, 2002, pp. 25–58.
27. *Moore v. Regents*. University of California (Ca 1990).
28. RK Seide, CL Stephens. Ethical issues and application of patent laws in biotechnology. In *Who Owns Life*? D Magnus, A Caplan, G McGee, Eds. New York: Prometheus Books, 2002, pp. 59–73.
29. RK Seide, CL Stephens. Ethical issues and application of patent laws in biotechnology. In *Who Owns Life*? D Magnus, A Caplan, G McGee, Eds. New York: Prometheus Books, 2002, pp. 59–73.
30. T Caulfield, ER Gold, MK Cho. Patenting human genetic material: refocusing the debate. *Nat Rev Genet* 1:227–231, 2000.
31. Universal Declaration on the Human Genome and Human Rights. Paris: UNESCO, International Bioethics Committee, 1997.
32. Parlimentary Assembly of the Council of Europe. Recommendation 1425 (1999) and Recommendation 1468 (2000).
33. JA Goldstein, E Golod. Human gene patents. *Acad Med* 77:1315–1328, 2002.
34. BM Knoppers. Status, sale and patenting of human genetic material: an international survey. *Nat Genet* 22:23–26, 1999.
35. D Nelkin, L Andrews. Homo economicus: commercialization of body tissue in the age of biotechnology. *Hastings Center Rep* 28:30–39, 1998.
36. PN Ossorio. Property rights and human bodies. In *Who Owns Life*? D Magnus, A Caplan, G McGee, Eds. New York: Prometheus Books, 2002, pp. 223–242.
37. D Nelkin, L Andrews. Homo economicus: commercialization of body tissue in the age of biotechnology. *Hastings Center Rep* 28:30–39, 1998.
38. JF Merz, MK Cho, MJ Robertson, DGB Leonard. Disease gene patenting is a bad innovation. *Mol Diagn* 2:299–304, 1997.
39. JA Goldstein, E Golod. Human gene patents. *Acad Med* 77:1315–1328, 2002.
40. JF Merz, MK Cho, MJ Robertson, DGB Leonard. Disease gene patenting is a bad innovation. *Mol Diagn* 2:299–304, 1997.
41. SM Thomas, MM Hopkins, M Brady. Shares in the human genome: the future of patenting DNA. *Nat Biotechnol* 20:1185–1188, 2002.
42. JA Goldstein, E Golod. Human gene patents. *Acad Med* 77:1315–1328, 2002.
43. GC Elliott. A brief guide to understanding patentability and the meaning of patents. *Acad Med* 77:1309–1314, 2002.
44. JA Goldstein, E Golod. Human gene patents. *Acad Med* 77:1315–1328, 2002.
45. FM Scherer. The economics of human gene patents. *Acad Med* 77:1348–1367, 2002.

46. KI Kaitin, NR Bryant, L Lasagna. The role of the research-based pharmaceutical industry in medical progress in the United States. *J Clin Pharmacol* 33:412–417, 1993.
47. Pharmaceutical Research and Manufacturers of America. Industry Profile 2003. Washington DC: PhRMA, 2003. Available at www.phrma.org/publications/publications/profile02/index.cfm (accessed July 31, 2003).
48. JA Goldstein, E Golod. Human gene patents. *Acad Med* 77:1315–1328, 2002.
49. SM Thomas, MM Hopkins, M Brady. Shares in the human genome: the future of patenting DNA. *Nat Biotechnol* 20:1185–1188, 2002.
50. GC Elliott. A brief guide to understanding patentability and the meaning of patents. *Acad Med* 77:1309–1314, 2002.
51. L Bendekgey, D Hamlet-Cox. Gene patents and innovation. *Acad Med* 77:1373–1380, 2002.
52. RS Eisenberg. Why the gene patenting controversy persists. *Acad Med* 77:1381–1387, 2002.
53. RS Eisenberg. Genomics in the public domain: strategy and policy. *Nat Rev Genet* 1:70–74, 2000.
54. SM Thomas, MM Hopkins, M Brady. Shares in the human genome: the future of patenting DNA. *Nat Biotechnol* 20:1185–1188, 2002.
55. MA Heller, RS Eisenberg. Can patents deter innovation? The anticommons in biomedical research. *Science* 280:698–701, 1998.
56. A Schissel, JF Merz, MK Cho. Survey confirms fears about licensing of genetic tests. *Nature* 402:118, 1999.
57. D Magnus. Disease gene patenting: the clinician's dilemma. *Cambridge Q Healthcare Ethics* 7:433–435, 1998.
58. Nuffield Council on Bioethics. *The Ethics of Patenting DNA*. London: Nuffield Council on Bioethics, 2002. 58a. *USPTO Utility Examination Guidelines Fed Reg*, Vol. 66, p. 1092, January 5, 2001.
59. L Bendekgey, D Hamlet-Cox. Gene patents and innovation. *Acad Med* 77:1373–1380, 2002.
60. T Caulfield, ER Gold, MK Cho. Patenting human genetic material: refocusing the debate. *Nat Rev Genet* 1:227–231, 2000.
61. T Caulfield, ER Gold, MK Cho. Patenting human genetic material: refocusing the debate. *Nat Rev Genet* 1:227–231, 2000.
62. GC Elliott. A brief guide to understanding patentability and the meaning of patents. *Acad Med* 77:1309–1314, 2002.
63. T Caulfield, ER Gold, MK Cho. Patenting human genetic material: refocusing the debate. *Nat Rev Genet* 1:227–231, 2000.
64. JH Barton. Intellectual property rights: reforming the patent system. *Science* 287:1933, 2000.
65. T Caulfield, ER Gold, MK Cho. Patenting human genetic material: refocusing the debate. *Nat Rev Genet* 1:227–231, 2000.
66. D Blumenthal. Growing pains for new academic/industry relationships. *Health Aff* 13:176–193, 1994.
67. D Blumenthal. Growing pains for new academic/industry relationships. *Health Aff* 13:176–193, 1994.
68. Bayh-Dole Act. Pub. L. 96-517.
69. Federal Technology Transfer Act. Pub. L. 99-502.
70. *Diamond v. Chakarbarty.* 447 US 303 (1980).

71. AC Gelijns, SO Their. Medical innovation and institutional interdependence. Rethinking university-industry connections. *JAMA* 287:72–77, 2002.
72. JE Bekelman, Y Li, CP Gross. Scope and impact of financial conflicts of interest in biomedical research. A systematic review. *JAMA* 289:454–465, 2003.
73. JE Bekelman, Y Li, CP Gross. Scope and impact of financial conflicts of interest in biomedical research. A systematic review. *JAMA* 289:454–465, 2003.
74. D Blumenthal. Growing pains for new academic/industry relationships. *Health Aff* 13:176–193, 1994.
75. The Association of University Technology Managers. The AUTM Licensing Survey: FY 2001. Available at http://www.autm.net (accessed July 30, 2003).
76. D Blumenthal. Growing pains for new academic/industry relationships. *Health Aff* 13:176–193, 1994.
77. P Vallance. Biotechnology and new companies arising from academia. *Lancet* 358:1804–1806, 2001.
78. La Montagne. Biotechnology and research: promise and problems. *Lancet* 358:1723–1724, 2002.
79. H Moses, E Braunwald, JB Martin, SO Their. Collaborating with industry: choices for the academic medical center. *N Engl J Med* 347:1371–1375, 2002.
80. D Blumenthal, N Causino, E Campbell, et al. Relationships between academic institutions and industry in the life sciences: an industry survey. *N Engl J Med* 334:368–373, 1996.
81. D Blumenthal, N Causino, E Campbell, et al. Relationships between academic institutions and industry in the life sciences: an industry survey. *N Engl J Med* 334:368–373, 1996.
82. D Blumenthal, N Causino, E Campbell, et al. Relationships between academic institutions and industry in the life sciences: an industry survey. *N Engl J Med* 334:368–373, 1996.
83. D Weatherall. Academia and industry: increasingly uneasy bedfellows. *Lancet* 355:1574, 2000.
84. D Korn. Conflicts of interest in biomedical research. *JAMA* 284:2234–2237, 2000.
85. D Blumenthal. Growing pains for new academic/industry relationships. *Health Aff* 13:176–193, 1994.
86. D Blumenthal. Growing pains for new academic/industry relationships. *Health Aff* 13:176–193, 1994.
87. H Moses, E Braunwald, JB Martin, SO Their. Collaborating with industry: choices for the academic medical center. *N Engl J Med* 347:1371–1375, 2002.
88. H Moses, E Braunwald, JB Martin, SO Their. Collaborating with industry: choices for the academic medical center. *N Engl J Med* 347:1371–1375, 2002.
89. AC Gelijns, SO Their. Medical innovation and institutional interdependence. Rethinking university-industry connections. *JAMA* 287:72–77, 2002.
90. JB Martin, DL Kasper. In whose best interest? Breaching the academic-industrial wall. *N Engl J Med* 343:1646–1649, 2000.
91. D Blumenthal. Growing pains for new academic/industry relationships. *Health Aff* 13:176–193, 1994.
92. JB Martin, DL Kasper. In whose best interest? Breaching the academic-industrial wall. *N Engl J Med* 343:1646–1649, 2000.
93. AC Gelijns, SO Their. Medical innovation and institutional interdependence. Rethinking university-industry connections. *JAMA* 287:72–77, 2002.
94. AC Gelijns, SO Their. Medical innovation and institutional interdependence. Rethinking university-industry connections. *JAMA* 287:72–77, 2002.

95. D Kennedy. Enclosing the research commons. *Science* 294:2249, 2001.
96. D Kennedy. Enclosing the research commons. *Science* 294:2249, 2001.
97. JB Martin, DL Kasper. In whose best interest? Breaching the academic-industrial wall. *N Engl J Med* 343:1646–1649, 2000.
98. JE Bekelman, Y Li, CP Gross. Scope and impact of financial conflicts of interest in biomedical research. A systematic review. *JAMA* 289:454–465, 2003.
99. D Korn. Conflicts of interest in biomedical research. *JAMA* 284:2234–2237, 2000.
100. M Angell. Is academic medicine for sale? *N Engl J Med* 342:1516–1518, 2000.
101. DF Thompson. Understanding financial conflicts of interest. *N Engl J Med* 329:573–576, 1993.
102. MK Cho, R Shohara, A Schissel, D Rennie. Policies on faculty conflicts of interest at U.S. universities. *JAMA* 284:2203–2208, 2000.
103. D Korn. Conflicts of interest in biomedical research. *JAMA* 284:2234–2237, 2000.
104. JE Bekelman, Y Li, CP Gross. Scope and impact of financial conflicts of interest in biomedical research. A systematic review. *JAMA* 289:454–465, 2003.
105. H Moses, JB Martin. Academic relationships with industry. A new model for biomedical research. *JAMA* 285:933–935, 2001.
106. H Moses, JB Martin. Academic relationships with industry. A new model for biomedical research. *JAMA* 285:933–935, 2001.
107. D Korn. Conflicts of interest in biomedical research. *JAMA* 284:2234–2237, 2000.
108. D Blumenthal, EG Campbell, H Causino, KS Louis. Participation in life-science faculty in research relationships with industry. *N Engl J Med* 335:1734–1739, 1996.
109. D Blumenthal, N Causino, E Campbell, et al. Relationships between academic institutions and industry in the life sciences: an industry survey. *N Engl J Med* 334:368–373, 1996.
110. D Blumenthal, N Causino, E Campbell, et al. Relationships between academic institutions and industry in the life sciences: an industry survey. *N Engl J Med* 334:368–373, 1996.
111. MK Cho, R Shohara, A Schissel, D Rennie. Policies on faculty conflicts of interest at U.S. universities. *JAMA* 284:2203–2208, 2000.
112. Pharmaceutical Research and Manufacturers of America. *Industry Profile 2003*. Washington, DC: PhRMA, 2003. Available at www.phrma.org/publications/publications/profile02/index.cfm (accessed July 31, 2003).
113. J Lexchin, LA Bero, B Djulbegovic, O Clark. Pharmaceutical industry sponsorship and research outcome and quality: systematic review. *Brit Med J* 326 (400): 1167–1170, May 31, 2003.
114. J Lexchin, LA Bero, B Djulbegovic, O Clark. Pharmaceutical industry sponsorship and research outcome and quality: systematic review. *Brit Med J* 326 (400): 1167–1170, May 31, 2003.
115. G Antes, I Chalmers. Under-reporting of clinical trials is unethical. *Lancet* 361:978–979, 2003.
116. B Djulbegovic, M Lacevic, A Cantor, et al. The uncertainty principle and industry-sponsored research. *Lancet* 356:635–638, 2000.
117. D Blumenthal. Growing pains for new academic/industry relationships. *Health Aff* 13:176–193, 1994.
118. JE Bekelman, Y Li, CP Gross. Scope and impact of financial conflicts of interest in biomedical research. A systematic review. *JAMA* 289:454–465, 2003.
119. T Bodenheimer. Uneasy alliance. Clinical investigators and the pharmaceutical industry. *N Engl J Med* 342:1539–1544, 2000.

120. D Blumenthal, EG Campbell, H Causino, KS Louis. Participation in life-science faculty in research relationships with industry. *N Engl J Med* 335:1734–1739, 1996.
121. A Schieppati, G Remuzzi, S Garattini. Modulating the profit motive to meet need of the less-developed world. *Lancet* 358:1638–1641, 2001.
122. D Blumenthal. Growing pains for new academic/industry relationships. *Health Aff* 13:176–193, 1994.
123. M Angell. Is academic medicine for sale? *N Engl J Med* 342:1516–1518, 2000.
124. D Blumenthal. Growing pains for new academic/industry relationships. *Health Aff* 13:176–193, 1994.
125. CD DeAngelis. Conflict of interest and the public trust. *JAMA* 284:2237–2238, 2000.
126. D Blumenthal. Growing pains for new academic/industry relationships. *Health Aff* 13:176–193, 1994.
127. AC Gelijns, SO Their. Medical innovation and institutional interdependence. Rethinking university-industry connections. *JAMA* 287:72–77, 2002.
128. S Krimsky, LS Rothenberg. Financial interest and its disclosure in scientific publications. *JAMA* 280:225–226, 1998.
129. SV McCrary, CB Anderson, J Jakovljevic, et al. A national survey of policies on disclosure of conflicts of interest in biomedical research. *N Engl J Med* 343:1621–1626, 2000.
130. Objectivity in Research, 60 *Federal Register* 35810 (1995) (codified at 42 CFR §50).
131. MK Cho, R Shohara, A Schissel, D Rennie. Policies on faculty conflicts of interest at U.S. universities. *JAMA* 284:2203–2208, 2000.
132. MME Johns, M Barnes, PS Florencio. Restoring balance to industry-academia relationships in an era of institutional financial conflicts of interest. Promoting research while maintaining trust. *JAMA* 289:741–746, 2003.
133. JB Martin, DL Kasper. In whose best interest? Breaching the academic-industrial wall. *N Engl J Med* 343:1646–1649, 2000.
134. SV McCrary, CB Anderson, J Jakovljevic, et al. A national survey of policies on disclosure of conflicts of interest in biomedical research. *N Engl J Med* 343:1621–1626, 2000.
135. H Moses, JB Martin. Academic relationships with industry. A new model for biomedical research. *JAMA* 285:933–935, 2001.
136. H Moses, JB Martin. Academic relationships with industry. A new model for biomedical research. *JAMA* 285:933–935, 2001.
137. MME Johns, M Barnes, PS Florencio. Restoring balance to industry-academia relationships in an era of institutional financial conflicts of interest. Promoting research while maintaining trust. *JAMA* 289:741–746, 2003.
138. MME Johns, M Barnes, PS Florencio. Restoring balance to industry-academia relationships in an era of institutional financial conflicts of interest. Promoting research while maintaining trust. *JAMA* 289:741–746, 2003.
139. International Committee of Medical Journal Editors. Uniform Requirements for Manuscripts Submitted to Biomedical Journals. 2001. Available at http://www.icmje.org (last accessed July 31, 2003).
140. F Davidoff, CD DeAngelis, JM Drazen, et al. Sponsorship, authorship, and accountability. *Lancet* 358:854–856, 2001.
141. D Shalala. Protecting human subjects: what must be done. *N Engl J Med* 343:808–810, 2000.

142. H Moses, E Braunwald, JB Martin, SO Their. Collaborating with industry: choices for the academic medical center. *N Engl J Med* 347:1371–1375, 2002.
143. JB Martin, DL Kasper. In whose best interest? Breaching the academic-industrial wall. *N Engl J Med* 343:1646–1649, 2000.
144. D Korn. Conflicts of interest in biomedical research. *JAMA* 284:2234–2237, 2000.
145. AC Gelijns, SO Their. Medical innovation and institutional interdependence. Rethinking university-industry connections. *JAMA* 287:72–77, 2002.
146. H Moses, E Braunwald, JB Martin, SO Their. Collaborating with industry: choices for the academic medical center. *N Engl J Med* 347:1371–1375, 2002.
147. JB Martin, DL Kasper. In whose best interest? Breaching the academic-industrial wall. *N Engl J Med* 343:1646–1649, 2000.
148. H Moses, JB Martin. Academic relationships with industry. A new model for biomedical research. *JAMA* 285:933–935, 2001.
149. MME Johns, M Barnes, PS Florencio. Restoring balance to industry-academia relationships in an era of institutional financial conflicts of interest. Promoting research while maintaining trust. *JAMA* 289:741–746, 2003.
150. JS Alpert. Conflicts of interest. Science, money and health. *Arch Intern Med* 162:535–637, 2002.
151. D Dickersin, D Rennie. Registering clinical trials. *JAMA* 290:516–523, 2003.
152. JE Bekelman, Y Li, CP Gross. Scope and impact of financial conflicts of interest in biomedical research. A systematic review. *JAMA* 289:454–465, 2003.
153. MK Cho, R Shohara, Schissel A, D Rennie. Policies on faculty conflicts of interest at U.S. universities. *JAMA* 284:2203–2208, 2000.
154. CP Gross, AR Gupta, HM Krumholz. Disclosure of financial competing interests in randomized controlled trials: cross sectional review. *BMJ* 326:526–527, 2003.
155. A Hussain, R Smith. Declaring financial competing interests: survey of five general medical journals. *BMJ* 323:263–264, 2001.
156. KA Schulman, DM Seils, JW Timbie, et al. A national survey of provisions in clinical-trial agreements between medical schools and industry sponsors. *N Engl J Med* 347:1135–1141, 2002.
157. C Holden. Conflict of interest: NEJM admits breaking its own tough rules. *Science* 287:1573, 2000.
158. B Woodward. Challenges to human subject protections in U.S. medical research. *JAMA* 282:1947–1952, 1999.
159. J Savulescu, I Chalmers, J Blunt. Are research ethics committees behaving unethically? Some suggestions for improving performance and accountability. *BMJ* 313:1390–1393, 1998.
160. R Horton. Conflict of interest in clinical research: opprobrium or obsession? *Lancet* 349:1112–1113, 1997.
161. R Horton. Conflict of interest in clinical research: opprobrium or obsession? *Lancet* 349:1112–1113, 1997.
162. DF Thompson. Understanding financial conflicts of interest. *N Engl J Med* 329:573–576, 1993.
163. R Horton. Conflict of interest in clinical research: opprobrium or obsession? *Lancet* 349:1112–1113, 1997.
164. KJ Rothman, CI Cann. Judging words rather than authors. *Epidemiology* 8:223–225, 1997.
165. RP Kelch. Maintaining the public trust in clinical research. *N Engl J Med* 346:285–286, 2002.

166. JB Martin, DL Kasper. In whose best interest? Breaching the academic-industrial wall. *N Engl J Med* 343:1646–1649, 2000.
167. JB Martin, DL Kasper. In whose best interest? Breaching the academic-industrial wall. *N Engl J Med* 343:1646–1649, 2000.
168. TR Cech, JS Leonard. Conflicts of interest: moving beyond disclosure. *Science* 291:989, 2001.
169. F Davidoff, CD DeAngelis, JM Drazen, et al. Sponsorship, authorship, and accountability. *Lancet* 358:854–856, 2001.
170. D Dickersin, D Rennie. Registering clinical trials. *JAMA* 290:516–523, 2003.
171. D Moher, KF Schulz, D Altman, for the CONSORT group. The CONSORT Statement: Revised Recommendations for Improving the Quality of Reports of Parallel-Group Randomized Trials, 2005.
172. JS Pober, CS Neuhauser, JM Pober. Obstacles facing translational research in academic medical centers. *FASEB J* 15:2303–2313, 2001.
173. M Yarborough, RR Sharp. Restoring and preserving trust in biomedical research. *Acad Med* 77:8–14, 2002.
174. H Moses, E Braunwald, JB Martin, SO Their. Collaborating with industry: choices for the academic medical center. *N Engl J Med* 347:1371–1375, 2002.
175. H Moses, E Braunwald, JB Martin, SO Their. Collaborating with industry: choices for the academic medical center. *N Engl J Med* 347:1371–1375, 2002.
176. JE Bekelman, Y Li, CP Gross. Scope and impact of financial conflicts of interest in biomedical research. A systematic review. *JAMA* 289:454–465, 2003.
177. NS Sung, WF Crowley, M Genel, et al. Central challenges facing the national clinical research enterprise. *JAMA* 289:1278–1287, 2003.
178. CD DeAngelis. Conflict of interest and the public trust. *JAMA* 284:2237–2238, 2000.
179. Pharmaceutical Research and Manufacturers of America. *Industry Profile 2003*. Washington, DC: PhRMA, 2003. Available at www.phrma.org/publications/publications/profile02/index.cfm (accessed July 31, 2003).
180. AC Gelijns, SO Their. Medical innovation and institutional interdependence. Rethinking university-industry connections. *JAMA* 287:72–77, 2002.
181. NS Sung, WF Crowley, M Genel, et al. Central challenges facing the national clinical research enterprise. *JAMA* 289:1278–1287, 2003.
182. KH Rothenberg, SF Terry. Before it's too late: addressing fear of genetic information. *Science* 297:196–197, 2002.
183. D Nelkin, MS Lindee. *The DNA Mystique: The Gene as Cultural* Icon. New York: WH Freeman, 1995.
184. GJ Annas. Privacy rules for DNA databanks: protecting coded 'future diaries.' *JAMA* 270:2346–2350, 1993.
185. PA Roche, GJ Annas. Protecting genetic privacy. *Nat Rev Genet* 2:392–396, 2001.
186. JL Hustead, J Goldman. Genetics and privacy. *Am J Law Med* 28:285–307, 2002.
187. PA Roche, GJ Annas. Protecting genetic privacy. *Nat Rev Genet* 2:392–396, 2001.
188. AL Allen. Genetic privacy: emerging concepts and values. In *Genetic Secrets: Protecting Privacy and Confidentiality in the Genetic Era*, MA Rothstein, Ed. New Haven, CT: Yale University Press, 1997, pp. 31–59.
189. AL Allen. Genetic privacy: emerging concepts and values. In *Genetic Secrets: Protecting Privacy and Confidentiality in the Genetic Era*, MA Rothstein, Ed. New Haven, CT: Yale University Press, 1997, pp. 31–59.

190. AL Allen. Genetic privacy: emerging concepts and values. In *Genetic Secrets: Protecting Privacy and Confidentiality in the Genetic Era*, MA Rothstein, Ed. New Haven, CT: Yale University Press, 1997, pp. 31–59.
191. MA Rothstein. Genetic privacy and confidentiality: why they are so hard to protect. *J Law Med Ethics* 26:198–204, 1998.
192. JL Hustead, J Goldman. Genetics and privacy. *Am J Law Med* 28:285–307, 2002.
193. PA Roche, GJ Annas. Protecting genetic privacy. *Nat Rev Genet* 2:392–396, 2001.
194. JL Hustead, J Goldman. Genetics and privacy. *Am J Law Med* 28:285–307, 2002.
195. MR Anderlik, MA Rothstein. Privacy and confidentiality of genetic information: what rules for the new science? *Annu Rev Genomics Hum Genet* 2:401–433, 2001.
196. KH Rothenberg, SF Terry. Before it's too late: addressing fear of genetic information. *Science* 297:196–197, 2002.
197. PA Roche, GJ Annas. Protecting genetic privacy. *Nat Rev Genet* 2:392–396, 2001.
198. MR Anderlik, MA Rothstein. Privacy and confidentiality of genetic information: what rules for the new science? *Annu Rev Genomics Hum Genet* 2:401–433, 2001.
199. EW Clayton. Ethical, legal, and social implications of genomic medicine. *N Engl J Med* 249:562–569, 2003.
200. PA Roche, GJ Annas. Protecting genetic privacy. *Nat Rev Genet* 2:392–396, 2001.
201. MJ Green, JR Botkin. "Genetic exceptionalism" in medicine: clarifying the differences between genetic and nongenetic tests. *Ann Intern Med* 138:571–575, 2003.
202. TH Murray. Genetic exceptionalism and "future diaries": is genetic information different from other medical information? In *Genetic Secrets Protecting Privacy and Confidentiality in the Genetic Era*, MA Rothstein, Ed. London: University Press, 1997, pp. 60–73.
203. TH Murray. Genetic exceptionalism and "future diaries": is genetic information different from other medical information? In *Genetic Secrets Protecting Privacy and Confidentiality in the Genetic Era*, MA Rothstein, Ed. London: University Press, 1997, pp. 60–73.
204. TH Murray. Genetic exceptionalism and "future diaries": is genetic information different from other medical information? In *Genetic Secrets Protecting Privacy and Confidentiality in the Genetic Era*, MA Rothstein, Ed. London: University Press, 1997, pp. 60–73.
205. ES Lander. Scientific commentary: the scientific foundations and medical and social prospects of the Human Genome Project. *J Law Med Ethics* 26:184–188, 1998.
206. MJ Green, JR Botkin. "Genetic exceptionalism" in medicine: clarifying the differences between genetic and nongenetic tests. *Ann Intern Med* 138:571–575, 2003.
207. Health Insurance Portability and Accountability Act. Pub. L. 104-191, 110 Stat. 1936 (1996).
208. GJ Annas. The limits of state laws to protect genetic information. *N Engl J Med* 345:385–388, 2001.
209. GJ Annas. The limits of state laws to protect genetic information. *N Engl J Med* 345:385–388, 2001.
210. National Conference of State Legislators. State Genetic Nondiscrimination in Health Insurance Laws. Available at http://www.ncsl.org/programs/health/genetics/ndishlth.htm (accessed August 8, 2003).

211. National Conference of State Legislators. State Genetic Privacy Laws. Available at http://www.ncsl.org/programs/health/genetics/prt.htm (accessed August 8, 2003).
212. National Conference of State Legislators. State Genetics Employment Laws. Available at http://www.ncsl.org/programs/health/genetics/ndiscrim.htm (accessed August 8, 2003).
213. J Beck, JS Alper. Reconsidering genetic antidiscrimination legislation. *J Law Med Ethics* 26:205–210, 1998.
214. W Nowlan. A rational view of insurance and genetic discrimination. *Science* 297:195–196, 2002.
215. MA Hall, SS Rich. Laws restricting health insurers' use of genetic information: impact on genetic discrimination. *Am J Hum Genet* 66:293–307, 2000.
216. MA Hall, SS Rich. Laws restricting health insurers' use of genetic information: impact on genetic discrimination. *Am J Hum Genet* 66:293–307, 2000.
217. MA Hall, SS Rich. Laws restricting health insurers' use of genetic information: impact on genetic discrimination. *Am J Hum Genet* 66:293–307, 2000.
218. J Stephenson. Genetic test information fears unfounded. *JAMA* 282:2197–2198, 1999.
219. MA Hall, SS Rich. Genetic privacy laws and patients' fear of discrimination by health insurers: the view from genetic counselors. *J Law Med Ethics* 28:245–257, 2000.
220. W Nowlan. A rational view of insurance and genetic discrimination. *Science* 297:195–196, 2002.
221. PS Miller. Genetic discrimination in the workplace. *J Law Med Ethics* 26:189–197, 1998.
222. *United States Equal Employment Opportunity Commission v. Burlington Northern and Santa Fe Railway.* Civ. C01-4013 MWB (N.D. Iowa 2001).
223. U.S. Congressional Office of Technology Assessment. Genetic Monitoring and Screening in the Workplace. 1990. Or The National Human Genome Research Institute. Genetic Information and the Workplace. Department of Labor, Department of Health and Human Services, Equal Employment Opportunity Commission, Department of Justice. Available at http://www.nhgri. nih.gov/HGP/Reports/genetics_workplace.html (accessed August 31, 2003).
224. JL Hustead, J Goldman. Genetics and privacy. *Am J Law Med* 28:285–307, 2002.
225. W Nowlan. A rational view of insurance and genetic discrimination. *Science* 297:195–196, 2002.
226. MA Hall, SS Rich. Laws restricting health insurers' use of genetic information: impact on genetic discrimination. *Am J Hum Genet* 66:293–307, 2000.
227. KH Rothenberg, SF Terry. Before it's too late: addressing fear of genetic information. *Science* 297:196–197, 2002.
228. PR Reilly. Rethinking risks to human subjects in genetic research. *Am J Hum Genet* 63:682–685, 1998.
229. PR Reilly. Rethinking risks to human subjects in genetic research. *Am J Hum Genet* 63:682–685, 1998.
230. TT Ashburn, SK Wilson, BI Eisenstein. Human tissue research in the genomic era of medicine. Balancing individual and societal interests. *Arch Intern Med* 160:3377–3384, 2000.
231. Federal Policy for the Protection of Human Subjects. 45 CFR (1991).
232. 21 CFR 50.25 (2002).

233. JF Merz. On the intersection of privacy, consent, commerce and genetics research. In *Populations and Genetics: Legal Socio-Ethical Perspectives*, BM Knoppers, Ed. New York: Kluwer Legal International, forthcoming, 2003.
234. TT Ashburn, SK Wilson, BI Eisenstein. Human tissue research in the genomic era of medicine. Balancing individual and societal interests. *Arch Intern Med* 160:3377–3384, 2000.
235. National Bioethics Advisory Commission. *Research Involving Human Biological Materials: Ethical Issues and Policy Guidance*, Vol 1. Rockville, MD: National Bioethics Advisory Commission, 1999. Available at http://bioethics.gov/pubs.html (accessed August 31, 2003).
236. EW Clayton, KK Steinberg, MJ Khoury, et al. Informed consent for genetic research on stored tissue samples. *JAMA* 274:1786–1792, 1995.
237. LM Beskow, W Burke, JF Merz, et al. Informed consent for population-based research involving genetics. *JAMA* 286:2315–2321, 2001.
238. MA Austin. Ethical issues in human genome epidemiology: a case study based on the Japanese American Family Study in Seattle, Washington. *Am J Epidemiol* 155:585–592, 2002.
239. JF Merz, P Sankar, SE Raube, V Livoisi. Use of human tissues in research: clarifying clinician and researcher roles and information flows. *J Invest Med* 45:252–257, 1997.
240. JF Merz, P Sankar, SE Raube, V Livoisi. Use of human tissues in research: clarifying clinician and researcher roles and information flows. *J Invest Med* 45:252–257, 1997.
241. BP Fuller, MJ Ellis Kahn, PA Barr, et al. Privacy in genetics research. *Science* 285:1359–1361, 1999.
242. R McKie. The gene collector. *BMJ* 321:854, 2000.
243. L Frank. Storm brews over gene bank of Estonian population. *Science* 286:1262–1263, 1999.
244. GJ Annas. Rules for research on human genetic variation: lessons from Iceland. *N Engl J Med* 342:1830–1833, 2000.
245. GJ Annas. Rules for research on human genetic variation: lessons from Iceland. *N Engl J Med* 342:1830–1833, 2000.
246. TT Ashburn, SK Wilson, BI Eisenstein. Human tissue research in the genomic era of medicine. Balancing individual and societal interests. *Arch Intern Med* 160:3377–3384, 2000.
247. Human Genome Organization (HUGO). Statement on Human Genomic Databases. London: Human Genome Organization, 2002). Available at http://www.hugo-international.org/hugo/HEC_Dec02.html (accessed August 31, 2003).
248. MW Foster, RR Sharp, WL Freeman, et al. The role of community review in evaluating the risks of human genetic variation research. *Am J Hum Genet* 64:1719–1727, 1999.
249. ET Juengst. Commentary: what "community review" can and cannot do. *J Law Med Ethics* 28:52–54, 2000.
250. RR Sharp, MW Foster. Involving study populations in the review of genetic research. *J Law Med Ethics* 28:41–51, 2000.
251. RR Sharp, MW Foster. Involving study populations in the review of genetic research. *J Law Med Ethics* 28:41–51, 2000.
252. L Andrews, D Nelkin. Whose body is it anyway? Disputes over body tissue in a biotechnology age. *Lancet* 351:53–57, 1998.

253. BP Fuller, MJ Ellis Kahn, PA Barr, et al. Privacy in genetics research. *Science* 285:1359–1361, 1999.
254. PA Roche, GJ Annas. Protecting genetic privacy. *Nat Rev Genet* 2:392–396, 2001.
255. BP Fuller, MJ Ellis Kahn, PA Barr, et al. Privacy in genetics research. *Science* 285:1359–1361, 1999.
256. E Marshall. Company plans to bank human DNA profiles. *Science* 291:575, 2001.
257. JF Merz, P Sankar, SE Raube, V Livoisi. Use of human tissues in research: clarifying clinician and researcher roles and information flows. *J Invest Med* 45:252–257, 1997.
258. PA Roche, GJ Annas. Protecting genetic privacy. *Nat Rev Genet* 2:392–396, 2001.
259. BP Fuller, MJ Ellis Kahn, PA Barr, et al. Privacy in genetics research. *Science* 285:1359–1361, 1999.
260. FS Collins. Shattuck lecture: medical and societal consequences of the Human Genome Project. *N Engl J Med* 341:28–37, 1999.
261. FS Collins, VA McKusick. Implications of the Human Genome Project for medical science. *JAMA* 285:540–544, 2001.
262. MJ Khoury, McCabe, ERB McCabe. Population screening in the age of genomic medicine. *N Engl J Med* 348:50–58, 2003.
263. ES Lander. Scientific commentary: the scientific foundations and medical and social prospects of the Human Genome Project. *J Law Med Ethics* 26:184–188, 1998.
264. RS Cooper, BM Psaty. Genomics and medicine: distraction, incremental progress, or the dawn of a new age? *Ann Intern Med* 138:576–580, 2003.
265. RS Cooper, BM Psaty. Genomics and medicine: distraction, incremental progress, or the dawn of a new age? *Ann Intern Med* 138:576–580, 2003.
266. MP Murphy. Current pharmacogenomic approaches to clinical drug development. *Pharmacogenomics* 1:115–123, 2000.
267. TF Bumol, AM Watanabe. Genetic information, genomic technologies, and the future of drug discovery. *JAMA* 285:551–555, 2001.
268. MP Murphy. Current pharmacogenomic approaches to clinical drug development. *Pharmacogenomics* 1:115–123, 2000.
269. AD Roses. Pharmacogenetics and future drug development and delivery. *Lancet* 355:1358–1361, 2000.
270. FS Collins. Genetics: an explosion of knowledge is transforming clinical practice. *Geriatrics* 54:41–47, 1999.
271. JM Rusnak, RM Kisabeth, DP Herbert, et al. Pharmacogenetics: a clinician's primer on emerging technologies for improved patient care. *Mayo Clin Proc* 76:299–309, 2001.
272. AD Roses. Pharmacogenetics. *Hum Mol Genet* 10:2261–2267, 2001.
273. JL Lazarou, BH Pomeranz, PN Corey. Incidence of adverse drug reactions in hospitalized patients. A meta-analysis of prospective studies. *JAMA* 279:1200–1205, 1998.
274. L Kohn, J Corrigan, M Donaldson, Eds. *To Err Is Human: Building a Safer Health System.* Washington, DC: Institute of Medicine, 2000.
275. FR Ernst, AJ Grizzle. Drug-related morbidity and mortality: updating the cost-of-illness model. *J Am Pharm Assoc* 41:192–199, 2001.
276. JM Rusnak, RM Kisabeth, DP Herbert, et al. Pharmacogenetics: a clinician's primer on emerging technologies for improved patient care. *Mayo Clin Proc* 76:299–309, 2001.

277. RS Cooper, BM Psaty. Genomics and medicine: distraction, incremental progress, or the dawn of a new age? *Ann Intern Med* 138:576–580, 2003.
278. NA Holtzman, TM Marteau. Will genetics revolutionize medicine? *N Engl J Med* 343:141–144, 2000.
279. NA Holtzman, TM Marteau. Will genetics revolutionize medicine? *N Engl J Med* 343:141–144, 2000.
280. NJ Wald, AK Hackshaw, CD Frost. When can a risk factor be used as a worthwhile screening test? *BMJ* 319:1562–1565, 1999.
281. NA Holtzman, TM Marteau. Will genetics revolutionize medicine? *N Engl J Med* 343:141–144, 2000.
282. TM Marteau, RT Croyle. Psychological responses to genetic testing. *BMJ* 316:693–696, 1998.
283. J Emery, S Hayflick. The challenge of integrating genetic medicine into primary care. *BMJ* 322:1027–1030, 2001.
284. W Burke, J Emery. Genetics education for primary-care providers. *Nat Rev Genet* 3:561–566, 2002.
285. FM Giardiello, JD Brensinger, GM Petersen, et al. The use and interpretations of commercial APC gene testing for familial adenomatous polyposis. *N Engl J Med* 336:823–827, 1997.
286. RS Cooper, BM Psaty. Genomics and medicine: distraction, incremental progress, or the dawn of a new age? *Ann Intern Med* 138:576–580, 2003.
287. T Walker. Fast forward. Predicting the future with genetic testing requires ethical forethought. *Manage Healthcare Exec* July 1:26–29, 2003.
288. JA Robertson, B Brody, A Buchanan, et al. Pharmacogenetic challenges for the health care system. *Health Aff* 21:155–167, 2002.
289. SL Burton, L Randel, K Titlow, et al. The ethics of pharmaceutical benefit management. *Health Aff* 20:150–163, 2001.
290. MT Lysaught. Commentary: reconstructing genetic research as research. *J Law Med Ethics* 26:48–54, 1998.
291. LR Churchill, ML Collins, NMP King, et al. Genetic research as therapy: implications of "gene therapy" for informed consent. *J Law Med Ethics* 26:38–47, 1998.
292. Robertson, B Brody, A Buchanan, et al. Pharmacogenetic challenges for the health care system. *Health Aff* 21:155–167, 2002.
293. MT Lysaught. Commentary: reconstructing genetic research as research. *J Law Med Ethics* 26:48–54, 1998.
294. LR Churchill, ML Collins, NMP King, et al. Genetic research as therapy: implications of "gene therapy" for informed consent. *J Law Med Ethics* 26:38–47, 1998.

Index

C

G

H

I

J

K

L

M

Q

R

200269571X